NATIONAL GEOGRAPHIC

# Explⓞring
# Science

## Program Consultants

Randy L. Bell, Ph.D.

Malcolm B. Butler, Ph.D.

Kathy Cabe Trundle, Ph.D.

Judith S. Lederman, Ph.D.

Center for the Advancement of Science in Space, Inc.

# Welcome to Exploring Science ....... 2

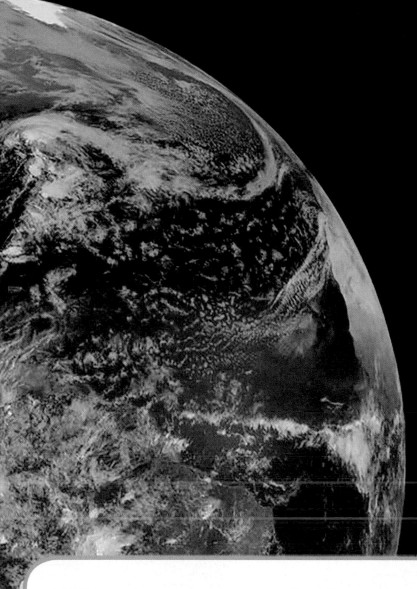

# Nature of Science ............ 8

# Life Science

## Structure, Function, and Information Processing

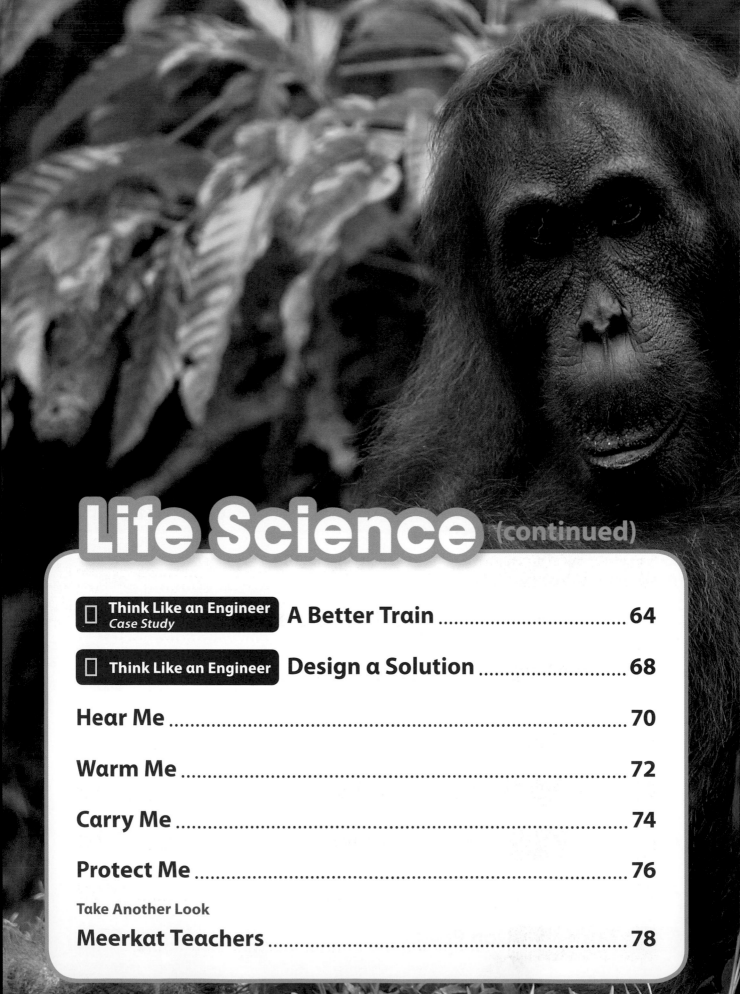

# Life Science (continued)

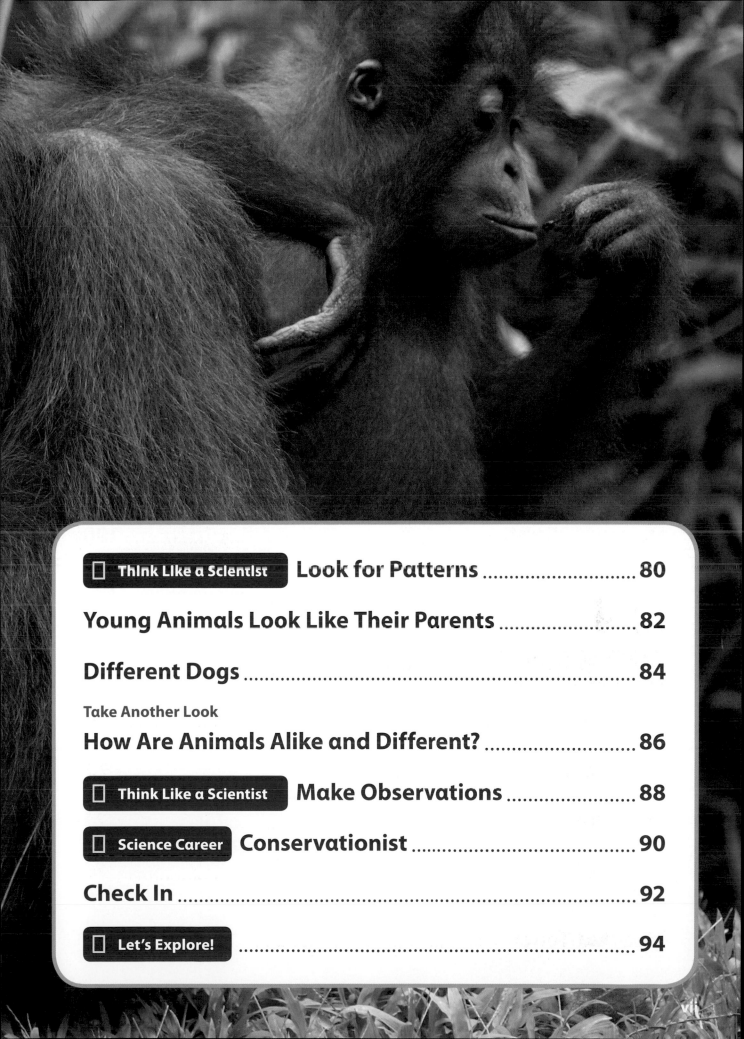

# Earth Science

## Space Systems: Patterns and Cycles

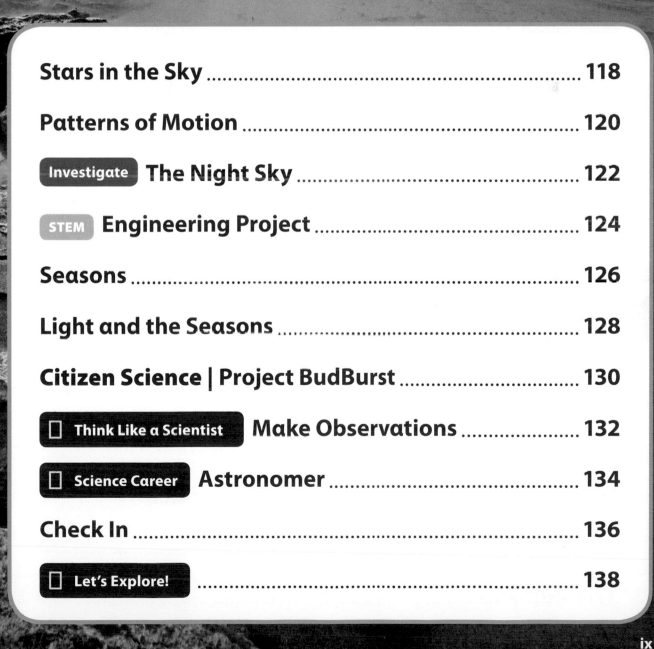

# Physical Science

## Waves: Light and Sound

**Nalini Nadkarni** is a forest ecologist. She studies living things in rain forests. Nalini's work helps protect rain forests.

Nalini investigates in a cloud forest. She is safe on a platform. It is high in the forest canopy.

# Welcome to Exploring Science!

Hello! I am Nalini Nadkarni. I am a forest ecologist. One place I study is the Monteverde Cloud Forest. It is in Costa Rica. A cloud forest has rain and fog. I climb to the top of the trees. This is called the forest canopy.

Many kinds of plants and animals live in the canopy. Some live nowhere else. I ask questions about them. I make observations. I record data. I look for patterns. I learn what the plants and animals need.

Rain forests are being destroyed. Many plants and animals may not survive. I encourage people to protect rain forests.

In *Exploring Science*, let's study how scientists learn about our world. You can act like a scientist, too. Are you ready? Let's begin!

# Keeping a
# Science Notebook 📓 My Science Notebook

I use a science notebook. I ask questions. I record observations and data. I look for patterns. I explain what I learn. You can keep a science notebook, too. Here are some ways to use your notebook.

- Make drawings of new science words and main ideas.
- Label drawings. Write to explain ideas.
- Collect photos, news stories, and other objects.
- Use tables, charts, or graphs to record explanations.
- Record reasons for explanations and conclusions.
- Write about what scientists and engineers do.
- Ask new questions.

Look at the notebook examples for some ideas. Now set up your science notebook!

A science notebook is one of a scientist's most important tools.

There are many types of living things in a cloud forest. Many live nowhere else.

▼ Use lists and pictures to show what you learned.

▼ Use drawings to show and tell what you observe in nature.

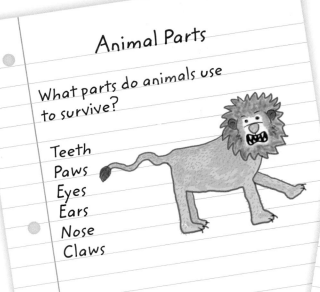

### Animal Parts

What parts do animals use to survive?

Teeth
Paws
Eyes
Ears
Nose
Claws

### Sun in the Sky

1
2
3

1. Sun in the morning
2. Sun at noon
3. Sun in the evening

▶ Write or draw observations in tables or charts.

### Sounds

| Object | Makes sound when plucked |
| --- | --- |
| string | yes |
| rubber band | yes |
| rope | no |
| shoelace | no |

# Set up Your Science Notebook

📓 My Science Notebook

Use your notebook every time you study science. Here are ways to get started. Your teacher may have more ideas. Use your own ideas, too.

- Make a cover. Draw something you like about science. Or draw something you would like to learn.

- In the front, write "Table of Contents." Leave blank pages.

- Add a number to each new page. Write the date.

- Use your notebook to see and share what you learn!

▼ Make a table of contents. Use your notebook every time you study science.

▼ Design a cover. Draw a science picture.

My Science Notebook

Table of Contents

| Date | Title | Page |
|------|-------|------|

# Why Explore Science?

I always loved to climb trees! I wondered about them. I grew up and learned about forests. Now I am a scientist. I ask questions about plants and animals. I study the rain forests. I tell people what I learn. I explain why it is important to protect rain forests.

You can act like a scientist, too. Think about how things in our world work. In *Nature of Science*, you will learn more about what science is and what it isn't. As you learn, write or draw in your science notebook!

Some plants are not on the ground. Nalini studies plants that grow on trees.

# Nature of Science

People always wondered about the sun, moon, and stars. They seemed to move in patterns. Look at the photo. How might the scene look an hour later?

# What Is Science?

Science is a way of knowing about our world. It is a way to explain what you observe. People **observed** the sun, moon, and stars. Long ago, they thought these objects moved around Earth.

Nicolaus Copernicus was a scientist. He was curious. He asked new questions. He observed how the sun, moon, and planets seemed to move. He used new observations and what he already knew to **infer.** He concluded that Earth and other planets moved around the sun. This was a new idea in science!

**DCI** ESS1.A: The Universe and Its Stars. Patterns of the motion of the sun, moon, and stars in the sky can be observed, described, and predicted. (1-ESS1-1)
**NS** Scientific Knowledge Is Based on Empirical Evidence. Scientists look for patterns and order when making observations about the world. (1-LS1-2)
**NS** Scientific Knowledge Assumes an Order and Consistency in Natural Systems. Science assumes natural events happen today as they happened in the past. (1-ESS1-1)
**NS** Scientific Knowledge Assumes an Order and Consistency in Natural Systems. Many events are repeated. (1-ESS1-1)

Scientists learned much more about Earth and the moon. This is a photo of Earth. It was taken by an astronaut on the moon.

Copernicus had to make sense out of his observations. Then he made a **model.** It explained how objects in space appeared to move. He drew Earth and other planets. They all moved around the sun.

# Wrap It Up!

1. List three things you observed today.

2. The sun, moon, and stars appear to move in the sky. How did Nicolaus Copernicus explain this?

# How Do Scientists Work?

In the 1960s, scientists **investigated** how to send people to the moon. They made detailed plans. They asked many new questions. They collected **data,** such as measurements and observations. They made sense out of the data. That was **evidence** to support their conclusions.

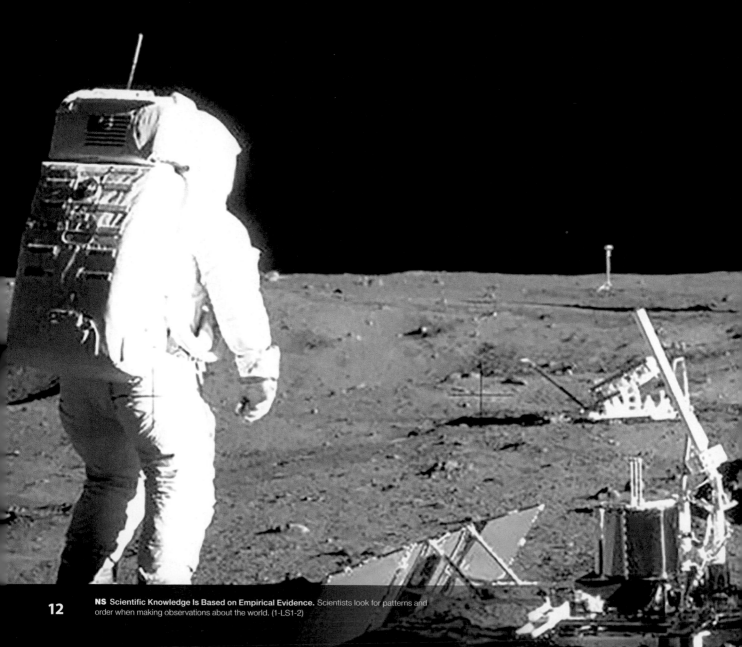

**NS** Scientific Knowledge Is Based on Empirical Evidence. Scientists look for patterns and order when making observations about the world. (1-LS1-2)

Scientists learned how far away the moon is. They learned how fast it moves. They learned its path. Scientists investigated how humans can live in space. Investigations lead to new questions. Science knowledge keeps growing. That's how science works!

How can people live in space? Scientists investigated. They made many tests. In a **fair test,** only one thing is changed.

## Wrap It Up! My Science Notebook

1. What do scientists do in investigations?

2. Why do you think it is important for scientists to keep asking new questions and doing new investigations?

# Who Are Scientists?

People of all backgrounds study science. Many scientists work in teams. Some do investigations in a space station. It moves in space around Earth!

Scientists are **curious**. They are creative. They ask new questions. They find new ways to investigate. Are you ready to think like a scientist? Then let's go!

Commander Peggy Whitson investigates. She is on the International Space Station. She studies samples of bone in very low gravity.

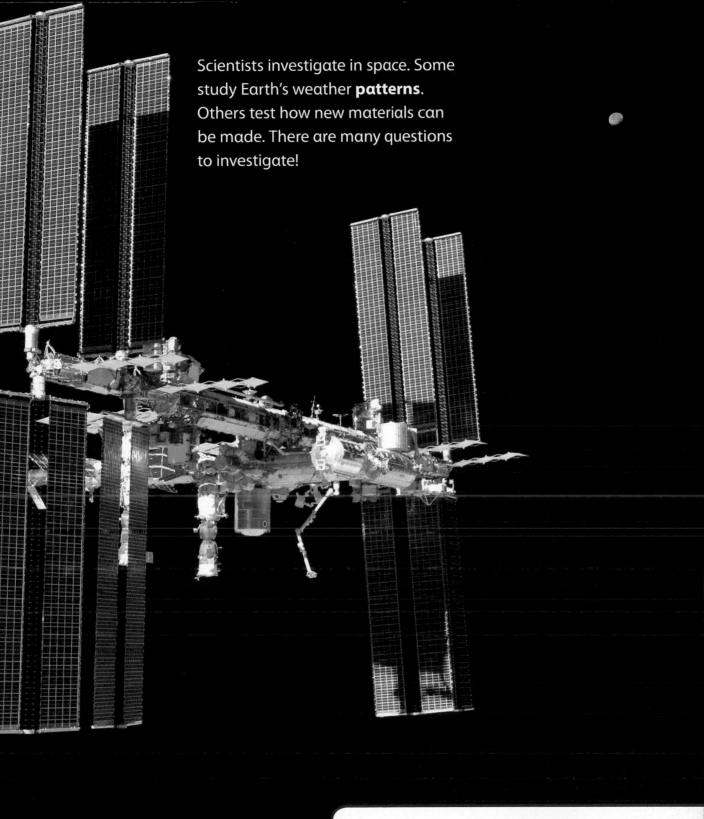

Scientists investigate in space. Some study Earth's weather **patterns**. Others test how new materials can be made. There are many questions to investigate!

## Wrap It Up!

1. Why are curiosity and creativity important to scientists?

2. Why do you think scientists often work in teams?

# Practice Science

**?** **How can you model how craters form?**

Chunks of rock crash on the moon. Craters form! Scientists use models to explain how things happen. You can make a model of craters forming.

Caption [TK]

## Materials

**pan of flour and cocoa powder**

**meterstick**

**bar magnet**

**magnetic marble**

**ruler**

**SEP Planning and Carrying Out Investigations.** Make observations (firsthand or from media) to collect data that can be used to make comparisons. (1-ESS1-2)
**SEP Analyzing and Interpreting Data.** Use observations (firsthand or from media) to describe patterns in the natural world in order to answer scientific questions. (1-ESS1-1)
**SEP Constructing Explanations and Designing Solutions.** Make observations (firsthand or from media) to construct an evidence-based account for natural phenomena. (1-LS3-1), (1-PS4-2)
**CCC Patterns.** Patterns in the natural and human designed world can be observed, used to describe phenomena, and used as evidence. (1-LS1-2), (1-LS3-1), (1-ESS1-1), (1-ESS1-2)
**NS Scientific Knowledge Assumes an Order and Consistency in Natural Systems.** Science assumes natural events happen today as they happened in the past. (1-ESS1-1)
**NS Scientific Knowledge Assumes an Order and Consistency in Natural Systems.** Many events are repeated. (1-ESS1-1)

**1** Drop a marble into the pan from 25 centimeters. Use the magnet to pick up the marble from the pan.

**2** Observe the model crater that formed. Measure the crater, and record your observations.

**3** Predict what will happen if you drop the marble from 50 centimeters. Record your prediction. Drop the marble, and remove it. Repeat Step 2.

**4** Ask a new question. Predict. Do your test. Record your observations.

## Wrap It Up!  My Science Notebook

1. How were the craters that formed in the model alike and different?

2. Did your observations support your predictions? What can you infer about how the marble hits the flour and powder from different heights?

3. How did you make sure you did a fair test? What was the only thing you changed in each trial?

17

**Nalini Nadkarni**  Forest Ecologist
National Geographic Explorer

# Let's Explore!

Scientists make observations. I observe how plants live and grow. I collect data. I look for patterns. I use new information and what I already know to infer. I make conclusions.

Life science is the study of living things. Topics include how plants and animals survive. Here are some questions you might investigate in *Life Science:*

- Why does a bald eagle have a hooked beak?
- How does hibernation help a dormouse survive?
- How can living in a group help meerkats survive?
- How is a young giraffe different from an adult giraffe?

Look at the notebook examples for ideas. As you read, think of your own questions. Look for answers. Then let's check in to see what you have learned!

## Parts of a Plant

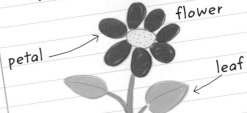

flower

petal

leaf

stem

roots

How do plant parts help the plant survive?

## My New Questions

1. What would happen if I grew a plant in the dark?

2. What gives a plant's leaves a green color?

3. If I put a plant in water with no soil, will it grow?

4. What are some plants that do not have flowers?

Young plants look like their parents. They look different too.

The young slash pine and the adult slash pine have thin, green needles and a trunk.

The trunk of the adult tree is tall and thick. The trunk of the young tree is short and thin. The adult tree is covered with thick bark.

The adult tree has pine cones with seeds. The young tree does not.

# Life Science

## Structure, Function, and Information Processing

King penguins head to the
ocean to find food.

# Plants

Plants are living things. Plants have different parts. The parts help the plant **survive,** or stay alive. The parts help the plant live and grow.

For example, leaves need sunlight to help a plant grow and survive. Many plants in shade have large leaves to collect more light. Many plants in sunny places have smaller leaves.

The vines on this tree are plants.

The trees are plants. Trees that grow tall can get plenty of sunlight. They may have smaller or thinner leaves.

This fern is a plant. It can grow in shady places. It has large leaves to collect light.

## Wrap It Up!

📖 My Science Notebook

1. Name three plants.

2. Some plants can grow in shade. Others grow in sunlight. How might the leaves of these plants be alike and different?

**DCI LS1.A: Structure and Function.** All organisms have external parts. Different animals use their body parts in different ways to see, hear, grasp objects, protect themselves, move from place to place, and seek, find, and take in food, water and air. Plants also have different parts (roots, stems, leaves, flowers, fruits) that help them survive and grow. (1-LS1-1)
**CCC Structure and Function.** The shape and stability of structures of natural and designed objects are related to their function(s). (1-LS1-1)

# Roots, Stems, and Leaves

How do you survive and grow? You use your body's parts to get what you need. Plants must get what they need, too. Read the captions to see how roots, stems, and leaves help this tree survive and grow.

Green plants use water, air, and light to make food in their **leaves.** Food moves from leaves to the stem and down to the roots.

The **stem** holds the plant upright. Stems hold up the leaves, too. Water and food move through the stem. The stem of a tree is called the trunk.

**Roots** take in water from the soil. Roots help hold the plant in place, too. Water moves from roots up to the leaves.

## Wrap It Up!

1. What do a plant's leaves do?

2. How do roots, stems, and leaves work together to help plants survive and grow?

**DCI** LS1.A: Structure and Function. All organisms have external parts. Different animals use their body parts in different ways to see, hear, grasp objects, protect themselves, move from place to place, and seek, find, and take in food, water and air. Plants also have different parts (roots, stems, leaves, flowers, fruits) that help them survive and grow. (1-LS1-1)
**ccc** Structure and Function. The shape and stability of structures of natural and designed objects are related to their function(s). (1-LS1-1)

25

# Flowers and Fruits

Many plants have flowers. Flowers come in many colors. They come in different shapes and sizes. Flowers can turn into fruits. The fruits contain seeds. New plants can grow from the seeds.

Cherries are the **fruit** of the cherry tree.

Cherries contain **seeds** that can grow into new cherry trees.

**Flowers** on the cherry tree can grow into cherries.

## Wrap It Up!

1. How can the flowers on a cherry tree change?

2. How do flowers help cherry trees survive and grow?

**DCI LS1.A: Structure and Function.** All organisms have external parts. Different animals use their body parts in different ways to see, hear, grasp objects, protect themselves, move from place to place and seek, find, and take in food, water and air. Plants also have different parts (roots, stems, leaves, flowers, fruits) that help them survive and grow. (1-LS1-1)
**CCC Structure and Function.** The shape and stability of structures of natural and designed objects are related to their function(s). (1-LS1-1)

# Plants and Light

**?** **What happens to a plant in a box?**

Plants need light to survive. In this activity, you will block some of the light a plant gets. You will see what happens to the plant.

## Materials

**bean plant in a pot**

**box with hole**

**DCI** LS1.D: Information Processing. Animals have body parts that capture and convey different kinds of information needed for growth and survival. Animals respond to these inputs with behaviors that help them survive. Plants also respond to some external inputs. (1-LS1-1)
**CCC** Structure and Function. The shape and stability of structures of natural and designed objects are related to their function(s). (1-LS1-1)

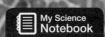

**1** Put a plant into a box that has a hole in it. Draw how the plant looks.

**2** Close the box. Place the box in a sunny place. The hole should face toward the sun.

**3** After one day, open the box. Look at the plant. Draw what you observe.

**4** Observe the plant every day for a week. Draw what you observe every day.

## Explore on Your Own

What would happen if you opened the box and observed the plant for another week? Make a plan to investigate. Carry out your plan. Record your observations. Compare the results of your investigations.

## Wrap It Up!  My Science Notebook

1. What did the plant look like after one day in the box? How did it look after one week?

2. What do you think caused the plant to grow this way?

# Root Growth

### ? How do roots grow?

You have seen how leaves and stems **respond,** or react, to light. Now you will investigate how roots respond.

## Materials

**tape**

**2 plastic cups**

**8 paper towels**

**2 bean seeds**

**spoon**

**water**

**ruler**

**clay**

**DCI LS1.D: Information Processing.** Animals have body parts that capture and convey different kinds of information needed for growth and survival. Animals respond to these inputs with behaviors that help them survive. Plants also respond to some external inputs. (1-LS1-1)
**CCC Structure and Function.** The shape and stability of structures of natural and designed objects are related to their function(s). (1-LS1-1)

**1** Label the cups. Put seeds and towels in the cups.

**2** Use the spoon to add water to the paper towels. Do this every other day. Watch for the seeds to sprout. Draw what you see.

**3** Wait for the roots to grow to 1.5 cm long. Put Cup A on its side. Use the clay to hold the cup in place.

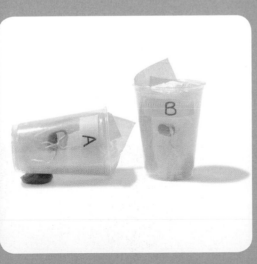

**4** Water the plants every other day. Observe how the roots grow. Record your observations.

## Wrap It Up!  My Science Notebook

1. How did the roots of the plants grow at first?

2. How did the root in Cup A change when the cup was on its side?

3. How do roots respond to a change in direction? How might this help the plant survive?

# Life Cycle of a Tomato Plant

Young plants can become adult plants. Adult plants can make new, young plants. The stages a plant goes through make up its **life cycle.** Follow the arrows in the picture. See how tomato plants can make young plants. A life cycle follows a **pattern.** It happens the same way again and again.

The tomatoes people eat are the fruit of the tomato plant.

**DCI** LS1.B: Growth and Development of Organisms. Adult plants and animals can have young. In many kinds of animals, parents and the offspring themselves engage in behaviors that help the offspring to survive. (1-LS1-2)
**CCC** Patterns. Patterns in the natural and human designed world can be observed, used to describe phenomena, and used as evidence. (1-LS1-2)

Each fruit holds many seeds.

Adult tomato plants can grow flowers. The flowers produce fruit.

A seed planted in soil can grow into a seedling.

**Seedlings** can grow into young plants. The young plants become adults.

## Wrap It Up! 📓 My Science Notebook

1. How is an adult tomato plant different from a seedling?

2. How do the parts of a seedling work together to help it survive?

3. What pattern can you observe in the life cycle of a tomato plant?

# Young Plants Look Like Their Parents

This is a young slash pine tree. It will grow to be a tall adult like the ones behind it. You can **compare** the young tree and the adults. You can tell how they are alike and different. The young tree and the adults both have skinny, green leaves called needles. They both have a stem, or trunk. Even though they are alike, they have some differences.

The young slash pine tree has soft needles. Its trunk is thin and bends easily.

**DCI** LS3.A: Inheritance of Traits. Young animals are very much, but not exactly like, their parents. Plants also are very much, but not exactly, like their parents. (1-LS3-1)
**CCC** Patterns. Patterns in the natural and human designed world can be observed, used to describe phenomena, and used as evidence. (1-LS3-1)

The needles have a waxy coating. This helps the plant survive in dry conditions.

The adult slash pine tree has a tall, thick trunk. It is covered with rough bark.

The adult slash pine tree has the same type of leaves as the young pine tree. This adult tree has pine cones that produce seeds. The young tree does not.

## Wrap It Up! My Science Notebook

1. How are the adult tree and the young tree alike?

2. How are the adult tree and the young tree different?

# Plants Can Be Different

You may think these are different kinds of plants. They are not. They are all adult zinnia plants. Flowers on zinnias grow in different sizes and colors. The flowers have different numbers of petals, too.

**DCI** LS3.B: **Variation of Traits.** Individuals of the same kind of plant or animal are recognizable as similar but can also vary in many ways. (1-LS3-1)
**CCC** **Patterns.** Patterns in the natural and human designed world can be observed, used to describe phenomena, and used as evidence. (1-LS3-1)

These zinnia plants all have different looking flowers. But their leaves all have the same shape.

## Wrap It Up! My Science Notebook

1. What might make you think that the picture shows different types of plants?

2. What might make you think that the plants in the picture are the same type?

# How Are Plants Alike and Different?

The young plant hasn't grown enough leaves yet to form a head of cabbage.

A head of cabbage grows in the center of the adult cabbage plant. It is a heavy ball made of layers of leaves.

**DCI** LS3.A: Inheritance of Traits. Young animals are very much, but not exactly like, their parents. Plants also are very much, but not exactly, like their parents. (1-LS3-1)
**DCI** LS3.B: Variation of Traits. Individuals of the same kind of plant or animal are recognizable as similar but can also vary in many ways. (1-LS3-1)
**CCC** Patterns. Patterns in the natural and human designed world can be observed, used to describe phenomena, and used as evidence. (1-LS3-1)

The adult lilac plant spreads into a bush as it grows.

The young lilac plant does not bloom yet.

## Share and Compare

- Choose a type of plant. On separate sheets of paper, draw one adult and one young of that plant.

- Mix up the pictures drawn by everyone in your group. Trade sets of pictures with another group.

- Match the pictures of adult plants with the young plants of the same kind.

- How do you know which pictures go together?

# Make Observations

You can make observations of adult and young plants. How are adult and young plants alike and different? How are the young plants alike and different from each other?

**1** **Plan an investigation.** 📓 My Science Notebook

Choose a type of plant. Decide how you will observe the plant. Will you use real plants or pictures? What will you look for? Draw or write out your plan in your science notebook.

**2** **Conduct an investigation.**

Observe details about the adult and young plants. Record your observations in your science notebook.

PE 1-LS3-1. Make observations to construct an evidence-based account that young plants and animals are like, but not exactly like, their parents.

The young oak tree is small. The few leaves use water, air, and light to make food for the plant. It has a thin stem. The roots are not deep, but they get water for the plant.

**3   Review your results.**

Review the details you have observed. Put the information in a chart that compares the young plant to the adult plant. How do your results answer the question?

**4   Share your results.**

Tell others about what you observed. Explain how your results show how young plants and adult plants are alike and different.

An adult oak can produce a fruit called an acorn. It has a hard covering that protects the seed inside.

An adult oak has a sturdy stem called a trunk. Thick bark protects living parts inside where water and food move. Deep roots take in water. They help to hold the huge tree up.

43

# Animal Parts

Animals are living things. Like plants, animals have different parts. The parts help animals survive and grow. Animals use their parts to get what they need to survive.

The caiman uses its parts to survive.

The caiman uses its tail to push it through the water.

It uses its webbed feet like paddles.

**DCI LS1.A: Structure and Function.** All organisms have external parts. Different animals use their body parts in different ways to see, hear, grasp objects, protect themselves, move from place to place, and seek, find, and take in food, water and air. Plants also have different parts (roots, stems, leaves, flowers, fruits) that help them survive and grow. (1-LS1-1)
**CCC Structure and Function.** The shape and stability of structures of natural and designed objects are related to their function(s). (1-LS1-1)

44

Its dark skin helps it hide in the mud.

It uses its strong jaws and sharp teeth to catch and eat other animals.

## Wrap It Up! My Science Notebook

1. What are some parts of the caiman?

2. How do a caiman's parts help it survive?

# Animals See and Hear

How do you find your way from home to school? You see where you are going. You hear what is happening around you. Animals see and hear, too. Some animals see and hear much better than people do!

Chameleons can move their eyes separately! They can see in every direction around them. You can't see a chameleon's ears, and it can't hear very well.

The ghost green crab can see all around itself, but not as clearly as you see. Crabs do not have ears. They feel sound with tiny hairs on their bodies.

**DCI** LS1.A: Structure and Function. All organisms have external parts. Different animals use their body parts in different ways to see, hear, grasp objects, protect themselves, move from place to place, and seek, find, and take in food, water and air. Plants also have different parts (roots, stems, leaves, flowers, fruits) that help them survive and grow. (1-LS1-1)
**CCC** Structure and Function. The shape and stability of structures of natural and designed objects are related to their function(s). (1-LS1-1)

Ring-tailed lemurs have big eyes and ears. They can see and hear very well.

## Wrap It Up! 📝 My Science Notebook

1. What parts allow these animals to see and hear?

2. Do all animals see and hear in the same way? Explain.

3. What do you use your eyes and ears to do? Why might animals need to see and hear?

47

# Animals Grasp

Could you spend a whole day with your hands in your pockets? That would be difficult! You need your hands to pick things up. Many animals do not have hands like yours. But they **grasp** things, too. They use other body parts to grasp things.

Birds grasp with their feet and their beaks. They can hold on tightly to branches. This bald eagle's toes have sharp claws.

African elephants can grasp with their trunks. They pick up food to place into their mouths.

**DCI LS1.A: Structure and Function.** All organisms have external parts. Different animals use their body parts in different ways to see, hear, grasp objects, protect themselves, move from place to place, and seek, find, and take in food, water and air. Plants also have different parts (roots, stems, leaves, flowers, fruits) that help them survive and grow. (1-LS1-1)
**CCC Structure and Function.** The shape and stability of structures of natural and designed objects are related to their function(s). (1-LS1-1)

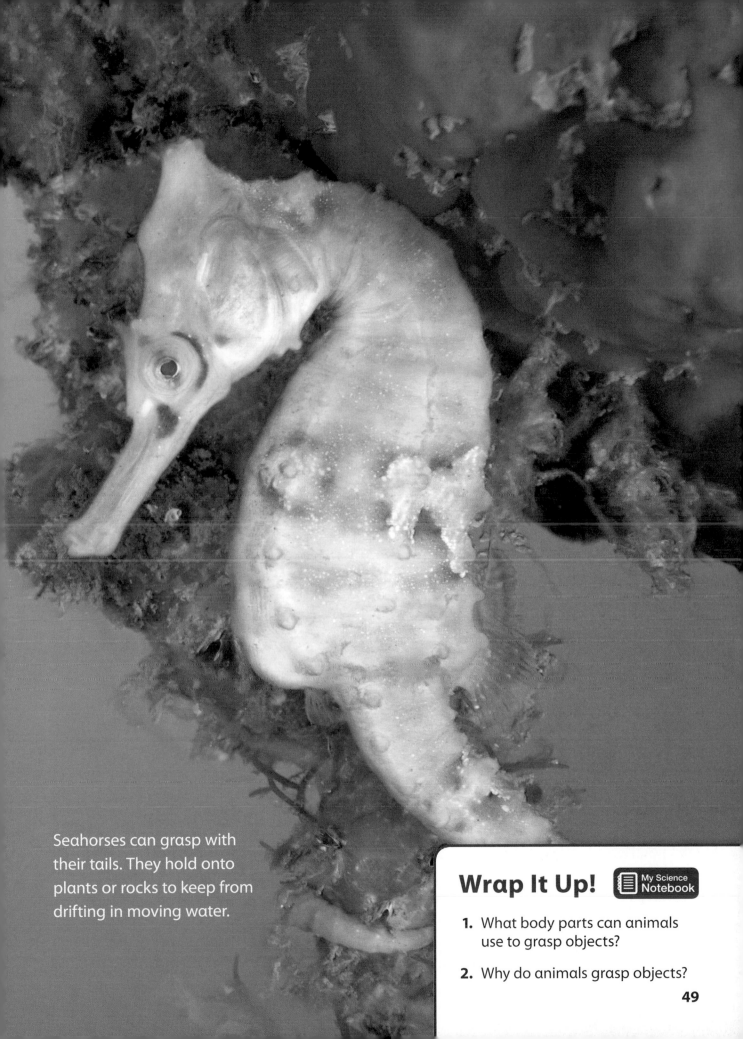

Seahorses can grasp with their tails. They hold onto plants or rocks to keep from drifting in moving water.

## Wrap It Up! 📓 My Science Notebook

1. What body parts can animals use to grasp objects?

2. Why do animals grasp objects?

# Animals Protect

Many animals eat other animals. Many animals need to **protect** themselves to keep from being eaten. They need to protect themselves from harsh weather, too. Animals use their body parts to protect themselves.

The leaf-tailed gecko looks like dried leaves and dirt. Its color protects it from other animals. How many geckos do you see?

**DCI LS1.A: Structure and Function.** All organisms have external parts. Different animals use their body parts in different ways to see, hear, grasp objects, protect themselves, move from place to place, and seek, find, and take in food, water and air. Plants also have different parts (roots, stems, leaves, flowers, fruits) that help them survive and grow. (1-LS1-1)
**CCC Structure and Function.** The shape and stability of structures of natural and designed objects are related to their function(s). (1-LS1-1)

The cape porcupine's sharp quills protect it from other animals. A sharp poke makes other animals leave it alone.

The Florida box turtle's hard shell protects its softer body parts. It can pull its head and legs inside.

## Wrap It Up! My Science Notebook

1. Why do many animals need protection?

2. The leaf-tailed gecko blends in with its home. Where does it probably live?

3. What does a gecko that lives high in a tree probably look like?

51

# Animals Move

Animals need to move around to survive. You use your legs and feet to move around. Like you, some animals walk and run. Others fly or swim. Animals use their body parts to move.

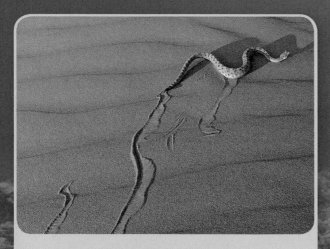

Snakes slither. This sidewinder rattlesnake bends from side to side. It uses its muscles to push itself forward.

The milkweed butterfly uses its wings to fly.

**DCI** LS1.A: Structure and Function. All organisms have external parts. Different animals use their body parts in different ways to see, hear, grasp objects, protect themselves, move from place to place, and seek, find, and take in food, water and air. Plants also have different parts (roots, stems, leaves, flowers, fruits) that help them survive and grow. (1-LS1-1)
**CCC** Structure and Function. The shape and stability of structures of natural and designed objects are related to their function(s). (1-LS1-1)

An octopus can move in two ways. It can crawl using its legs. It can also push water out of its body to move it forward.

## Wrap It Up! 📖 My Science Notebook

1. Name some body parts animals use to move.

2. Why do animals need to move?

# Animals Find What They Need

When you are hungry, you go to the kitchen to find a snack. Animals must find the food they need, too. Animals use their body parts to seek and find their food. The animals shown here eat other animals.

The nine-banded armadillo uses its claws to dig. It finds insects to eat. A shell made of bony plates protects the armadillo's body.

The Harris's hawk holds its prey with long, sharp claws. The claws are called talons.

**DCI** **LS1.A: Structure and Function.** All organisms have external parts. Different animals use their body parts in different ways to see, hear, grasp objects, protect themselves, move from place to place, and seek, find, and take in food, water and air. Plants also have different parts (roots, stems, leaves, flowers, fruits) that help them survive and grow. (1-LS1-1)
**CCC** **Structure and Function.** The shape and stability of structures of natural and designed objects are related to their function(s). (1-LS1-1)

A coyote has paws with pads. It hunts quietly so that it does not scare away prey.

## Wrap It Up! 📓 My Science Notebook

1. How are the paws of the coyote and the armadillo used differently?

2. Which of these animals seek other animals for food?

# Stories in Science

Some spiders spin webs.
An insect gets caught. The
spider feels vibrations. Then
it's dinnertime!

Some spider silk is sticky.
Insects get caught in it. Some
is strong. It can be used like
a safety line. There are more
kinds of spider silk.

**DCI LS1.A: Structure and Function.** All organisms have external parts. Different animals use their body parts in different ways to see, hear, grasp objects, protect themselves, move from place to place, and seek, find, and take in food, water and air. Plants also have different parts (roots, stems, leaves, flowers, fruits) that help them survive and grow. (1-LS1-1)

**DCI LS1.D: Information Processing.** Animals have body parts that capture and convey different kinds of information needed for growth and survival. Animals respond to these inputs with behaviors that help them survive. Plants also respond to some external inputs. (1-LS1-1)
**CCC Structure and Function.** The shape and stability of structures of natural and designed objects are related to their function(s). (1-LS1-1)

# Meet a Spider Woman

Cheryl Hayashi loves spiders. She got a job in college. She fed spiders. She now studies spider silk. It is strong and light.

A spider can make up to seven kinds of silk. It has a body part that makes each kind. Spiders use silk to wrap eggs. They use it to make webs. They use it to make traps. They use it to build nests. Hayashi hopes that people will learn to make spider silk. They may use it to make clothes, buildings, and other things.

Cheryl Hayashi asks questions. How can people make spider silk? How can they use it? She investigates to find answers.

## Wrap It Up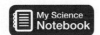

1. What are some ways that spiders use the silk they produce?

2. What are some ways that Cheryl Hayashi thinks people might use spider silk?

# Animals Take in Food, Water, and Air

Animals need food, water, and air. The animals shown here breathe air through their noses and mouths into their lungs. They eat food and drink water through their mouths.

Many animals on the savanna eat grasses. They have flat teeth for grinding the grasses. Some savanna animals eat other animals. They have sharp teeth for hunting and for eating meat.

Animals gather where there is water.

**DCI LS1.A: Structure and Function.** All organisms have external parts. Different animals use their body parts in different ways to see, hear, grasp objects, protect themselves, move from place to place, and seek, find, and take in food, water and air. Plants also have different parts (roots, stems, leaves, flowers, fruits) that help them survive and grow. (1-LS1-1)
**CCC Structure and Function.** The shape and stability of structures of natural and designed objects are related to their function(s). (1-LS1-1)

The lion's sharp teeth tear meat.

The elephant's flat teeth crush and grind plants.

## Wrap It Up! 📓 My Science Notebook

1. List three ways animals use their mouths.

2. How do the elephant's teeth and the lion's teeth differ?

# Animal Senses

The white-tailed deer is alert! It sees, hears, and smells a threat. When the deer senses danger, it flashes its tail. This lets other deer know to be on the lookout. Most times the deer will run away quickly.

Sometimes the deer will hold very still. Holding still makes the deer hard to see. It will wait until the danger passes before it moves again.

These deer hold very still as they listen to find out if danger is near.

**DCI LS1.D: Information Processing.** Animals have body parts that capture and convey different kinds of information needed for growth and survival. Animals respond to these inputs with behaviors that help them survive. Plants also respond to some external inputs. (1-LS1-1)
**CCC Structure and Function.** The shape and stability of structures of natural and designed objects are related to their function(s). (1-LS1-1)

White-tailed deer use their white tails to communicate. A flash says, "Danger is near!"

## Wrap It Up! My Science Notebook

1. What body parts does the deer use to sense what is going on around it?

2. How does the deer give information to other deer?

3. How does the deer respond to danger to help it survive?

# STEM
**SPACE STATION PROJECT**

SCIENCE
TECHNOLOGY
ENGINEERING
MATH

# Research Project: Eating Under Water and in Space

You know that animals use body parts to take in food. Now do a research project. Find out how some underwater animals take in food. Then learn how astronauts eat and drink in space. Astronauts train under water. It is something like being in space. Astronauts get used to eating and drinking in very low gravity.

**DCI** LS1.A: Structure and Function. All organisms have external parts. Different animals use their body parts in different ways to see, hear, grasp objects, protect themselves, move from place to place, and seek, find, and take in food, water and air. Plants also have different parts (roots, stems, leaves, flowers, fruits) that help them survive and grow. (1-LS1-1)
**SEP** Obtaining, Evaluating, and Communicating Information. Read grade-appropriate texts and use media to obtain scientific information to determine patterns in the natural world. (1-LS1-2)

It's mealtime on the International Space Station. The food is floating!

# The Challenge

Find out how an underwater animal gets food. Learn how astronauts eat and drink in space. Then share what you learn.

**1** **Select a topic.**

Choose an animal. It could be a baleen whale, a dolphin, a jellyfish, or a scallop. Research to answer questions. How does the animal find its food? How does it behave or use body parts to eat?

**2** **Plan and conduct research.**

Look for information about the animal. Your teacher will help you find sources. Take notes. Write a list of facts. Then you will learn how astronauts eat and drink in space.

**3** **Prepare your report.**

Make two posters. In one poster, show how the animal you chose finds and eats its food. In the other poster, show how astronauts eat in space. Draw and write examples.

**4** **Share.**

Work with your team. Prepare a presentation. Then share your posters with the class.

# A Better Train

## Problem

This train is the fastest train in the world. When the train comes out of a tunnel, it causes a startling noise. People did not like the noise. Eiji Nakatsu is the train's designer. He wanted to make the train quieter. He wondered if something in nature could help him solve this problem.

**DCI** ETS1.A: Defining and Delimiting Engineering Problems. Before beginning to design a solution, it is important to clearly understand the problem. (K–2-ETS1-1)
**CCC** Structure and Function. The shape and stability of structures of natural and designed objects are related to their function(s). (1-LS1-1)
**CETS** Influence of Engineering, Technology, and Science on Society and the Natural World. Every human-made product is designed by applying some knowledge of the natural world and is built using materials derived from the natural world. (1-LS1-1).

**Eiji Nakatsu** is an engineer. He works with a team. The team uses science to solve everyday problems.

This bullet train in Japan can travel over 320 kilometers (200 miles) per hour.

## Solution

Eiji Nakatsu likes to watch birds. He watched a kingfisher. The kingfisher can dive into water with very little splash. He designed the front of the train to look like the beak of a kingfisher. Now the train is quieter. It also is faster and uses less energy.

**Kingfisher**

**Bullet train**

Compare the shape of the train to the kingfisher's head and beak.

Kingfisher diving

## Wrap It Up! 📓 My Science Notebook

1. A bird moving into water can cause a splash. What can a train moving into open air from a tunnel cause?

2. How did Mr. Nakatsu use what he observed about the kingfisher to solve the problem of the loud train?

# Design a Solution

Animals and plants use their parts to protect themselves. People need to protect themselves, too. Like Eiji Nakatsu, you can look to nature for a solution.

**1**  **Define a problem.** My Science Notebook

How can you design a way of protecting yourself from danger?

**2**  **Design a solution.**

Think of a way that animals or plants protect themselves. Think of how you can protect yourself in a similar way. Draw a picture of your prototype in your science notebook. Explain what your prototype will do. Gather the materials you will need. Make your prototype.

The tortoise can pull its head and legs inside its hard shell.

**PE 1-LS1-1.** Use materials to design a solution to a human problem by mimicking how plants and/or animals use their external parts to help them survive, grow, and meet their needs.
**PE K-2-ETS1-1.** Ask questions, make observations, and gather information about a situation people want to change to define a simple problem that can be solved through the development of a new or improved object or tool.

**3** **Test and refine your solution.**
Test your prototype. Did it provide protection the way parts of an animal or a plant do? If not, change it. Draw your new design. Explain your changes.

**4** **Share your solution.**
Show others your prototype. Describe how your solution works. Explain how it is like the body parts an animal or a plant uses for protection. Explain how it is different.

# Hear Me

What happens when a baby cries? The baby's parents rush to see what the baby needs. Many kinds of baby animals cry, too. They make noises that let their parents know they need something. The parents react. They help the young animals get what they need to survive. Many kinds of young animals need help from their parents to survive.

The young grizzly bears make sounds to get their mother's attention. They might want to eat. They might want to play, too.

Lion cubs cry out to be fed their mother's milk. When they grow bigger the mother will quiet their hungry noises with meat from a hunt.

**DCI LS1.B: Growth and Development of Organisms.** Adult plants and animals can have young. In many kinds of animals, parents and the offspring themselves engage in behaviors that help the offspring to survive. (1-LS1-2)
**CCC Patterns.** Patterns in the natural and human designed world can be observed, used to describe phenomena, and used as evidence. (1-LS1-2)

The young birds are chirping loudly. They call for the mother cedar waxwing to feed them.

## Wrap It Up! 📓 My Science Notebook

1. Why are the young animals shown here making noises?

2. How do parent animals respond to help their young survive?

# Warm Me

Many young animals need help to keep warm. Many birds sit on their nests to warm their chicks.

Emperor penguins do not build nests. Young emperor penguins huddle close together. Some of the parents are away finding food. The young help keep each other warm.

**DCI** **LS1.B: Growth and Development of Organisms.** Adult plants and animals can have young. In many kinds of animals, parents and the offspring themselves engage in behaviors that help the offspring to survive. (1-LS1-2)

**CCC** **Patterns.** Patterns in the natural and human designed world can be observed, used to describe phenomena, and used as evidence. (1-LS1-2)

## Wrap It Up! 📓 My Science Notebook

1. Why do some adult animals warm their young?

2. How do some young animals stay warm without help from their parents?

# Carry Me

Animals move from place to place. They move to find food. They move to find shelter. Young animals cannot move as quickly as adults. Some parent animals hold onto their babies. At other times the babies do the holding on!

Young North American opossums grab onto their mother's fur and go for a ride.

**DCI** LS1.B: Growth and Development of Organisms. Adult plants and animals can have young. In many kinds of animals, parents and the offspring themselves engage in behaviors that help the offspring to survive. (1-LS1-2)
**CCC** Patterns. Patterns in the natural and human designed world can be observed, used to describe phenomena, and used as evidence. (1-LS1-2)

## Wrap It Up! 📓 My Science Notebook

1. What do young opossums do to help their mother carry them?

2. Why might a mother animal need to carry her young to a new place?

# Protect Me

The Arctic can be a dangerous place for a baby polar bear. The young cubs are small. Other animals hunt them for food. Mother polar bears carefully protect their young from the cold and from other animals.

Not all animal parents protect their young. Baby sea turtles hatch on beaches. They never see their mothers.

Polar bear cubs stay with their mothers for more than two years. Then they can live on their own.

**DCI** LS1.B: Growth and Development of Organisms. Adult plants and animals can have young. In many kinds of animals, parents and the offspring themselves engage in behaviors that help the offspring to survive. (1-LS1-2)
**CCC** Patterns. Patterns in the natural and human designed world can be observed, used to describe phenomena, and used as evidence. (1-LS1-2)

Baby sea turtles rush to the ocean as soon as they hatch. They are safer in the water.

## Wrap It Up! 📓 My Science Notebook

1. Do all animal parents take care of their young? Explain.

2. The Arctic is very cold. How might this mother bear be protecting her cub?

# Meerkat Teachers

What did you know when you were born? You needed to learn many things. Some animals are born knowing enough to survive. Other young animals must learn how to survive.

An adult teaches a young meerkat how to look for danger.

**DCI** LS1.B: Growth and Development of Organisms. Adult plants and animals can have young. In many kinds of animals, parents and the offspring themselves engage in behaviors that help the offspring to survive. (1-LS1-2)
**CCC** Patterns. Patterns in the natural and human designed world can be observed, used to describe phenomena, and used as evidence. (1-LS1-2)

Young meerkats must learn to find food. They do not know how until an adult teaches them.

## Share and Compare

- Think of the things that animals do to survive. Draw a picture of animals doing one of those things. Draw an adult and a young animal.

- Share your picture with others. What is the adult animal doing? What is the young animal doing? How do their behaviors help them to survive?

# Look for Patterns

Many animal parents help their young survive. Many young animals need food and protection. They also might need help moving from place to place.

Look at the pictures. Think about how each shows how parents help their young survive.

Mergansers

**PE 1-LS1-2.** Read texts and use media to determine patterns in behavior of parents and offspring that help offspring survive.

**Leopards**

**Polar bears**

**Giant petrels**

## Wrap It Up! 📖 My Science Notebook

1. Tell what is happening in each of these pictures.

2. How do the parent and young work together to help the young survive?

# Young Animals Look Like Their Parents

Human babies look very different from grown up people. They are smaller. They have a different body shape. They have less hair. Some adult and young animals look much the same. A mother giraffe and her young look very much alike. They are not exactly the same, though.

The young Masai giraffe will be the size of its mother after five to seven years.

## Wrap It Up!

1. How are the mother and young giraffe alike?

2. How are the mother and young giraffe different?

**DCI LS3.A: Inheritance of Traits.** Young animals are very much, but not exactly like, their parents. (1-LS3-1; Plants also are very much, but not exactly, like their parents. (1-LS3-1;
**CCC Patterns.** Patterns in the natural and human designed world can be observed, used to describe phenomena, and used as evidence. (1-LS3-1)

83

# Different Dogs

You can tell that all these animals are dogs. But look how different they are! They are different shapes, sizes, and colors. They also act in different ways. Some are calm. Others are very playful. Others are serious and make good guards.

This small dog has fluffy fur.

This larger dog has short hair.

**DCI** LS3.B: **Variation of Traits.** Individuals of the same kind of plant or animal are recognizable as similar but can also vary in many ways. (1-LS3-1)

**CCC** Patterns. Patterns in the natural and human designed world can be observed, used to describe phenomena, and used as evidence. (1-LS3-1)

## Wrap It Up! 📓 My Science Notebook

1. What do all dogs have in common?

2. What are some ways that these dogs are different from one another?

# How Are Animals Alike and Different?

A mother and young giraffe look very much alike. Other young animals differ more from their parents. Their size is one difference. You can see other differences as well.

All of these chicks have the same mother. How are the chicks alike and different from their mother? How are the chicks alike and different from each other?

**DCI** LS3.A: Inheritance of Traits. Young animals are very much, but not exactly like, their parents. Plants also are very much, but not exactly, like their parents. (1-LS3-1)
**DCI** LS3.B: Variation of Traits. Individuals of the same kind of plant or animal are recognizable as similar but can also vary in many ways. (1-LS3-1)
**CCC** Patterns. Patterns in the natural and human designed world can be observed, used to describe phenomena, and used as evidence. (1-LS3-1)

A mother harp seal and her pup

A white-tailed deer and her fawn

## Share and Compare

- Look at the animals in the pictures. How are the adults and young alike? How are the adults different from their young?

- Make a chart. Tell how parents and young are alike and different.

- Share your observations with your group. What did others observe?

# Make Observations

You can make observations to see how living things are the same and different. How are young animals and their parents alike and different? How are the young animals alike and different from each other?

**1** **Plan an investigation.** [My Science Notebook]

Look at many pictures of adult animals and young animals. Choose four types of animals. Record your observations in a chart.

**2** **Conduct an investigation.**

Write the name of an animal in your chart. Record one way that the adult and young are alike. Record one way they are different. Then record one way the young are alike. Record one way they are different from each other. Repeat for the other three animals you chose.

**PE** 1-LS3-1. Make observations to construct an evidence-based account that young plants and animals are like, but not exactly like, their parents.

### 3 Review your results.

Look at your chart. In what ways are the young alike and different from their parents? In what ways are the young alike and different from each other? Tell what you observed that supports your answers.

### 4 Share your results.

Share your chart with the class. Explain the results of your investigation.

These are wild pigs. The young are like, but not exactly like, their mother.

# Conservationist

To conserve something means to save it or protect it. A conservationist protects wildlife. Beverly and Dereck Joubert are conservationists.

The Jouberts explore Africa. They do research and make films. The Jouberts share what they learn with others. They hope other people will want to protect wildlife, too.

**NS** Scientific Investigations Use a Variety of Methods. Scientists use different ways to study the world. (1-PS4-1)
**NS** Scientific Investigations Use a Variety of Methods. Science investigations begin with a question. (1-PS4-1)

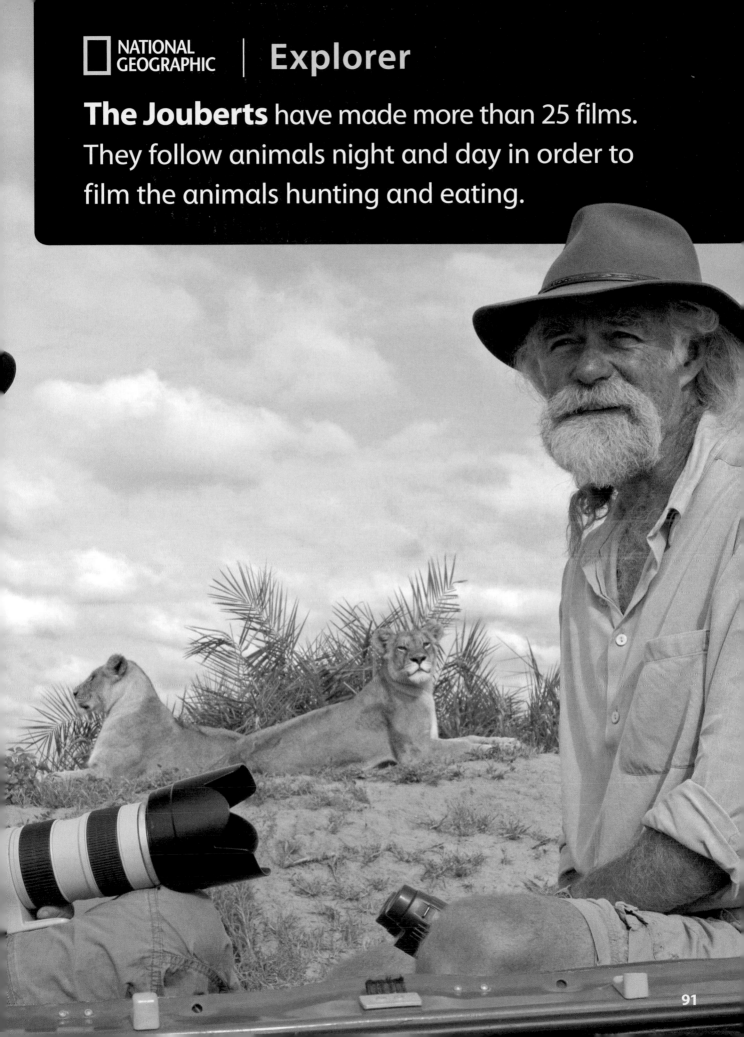

**The Jouberts** have made more than 25 films. They follow animals night and day in order to film the animals hunting and eating.

# Check In  My Science Notebook

Way to go! You have completed *Life Science.* Think about what you learned. Here is a checklist to help you check your progress. Look through your science notebook to find examples of the items. Tell how well you did. Write in your science notebook.

▼ Read the list. Think about what you recorded in your science notebook.

**For each item, select the choice that is true for you:  A. Yes  B. Not Yet**

- I made drawings of new science words and main ideas.

- I labeled drawings. I wrote to explain ideas.

- I collected photos, news stories, and other objects.

- I used tables, charts, or graphs to record observations.

- I recorded reasons for explanations and conclusions.

- I wrote about what scientists and engineers do.

- I asked new questions.

- I did something else. (Tell about it.)

## Reflect on Your Learning  My Science Notebook

1. Which investigation interested you the most? Explain why you liked it.

2. In what ways did you act like a scientist as you learned about life science?

Nalini shoots a line into the branches of a tree. Then she can attach ropes and cables. She is ready to climb the tree!

**Nalini Nadkarni**  Forest Ecologist
National Geographic Explorer

# Let's Explore!

Scientists are creative. I designed a tool. It shoots a line into tree branches. Then I use ropes. I climb to the canopy.

Earth science is the study of Earth and space. Some scientists study how the sun, moon, and stars appear to move. They study patterns of seasons. They study land, air, water, and soil. Here are some questions to answer in *Earth Science:*

- The moon seems to move in what pattern in the sky?

- In what ways do stars appear to move at night?

- Which season has the most hours of daylight? Which season has the fewest hours of daylight?

- What does an astronomer do?

Look at the notebook examples for ideas. Let's check in again later to review what you have learned!

▼ Record your observations and predictions when you investigate.

What can you observe about the position of the sun?

1. It is morning. I observe the sun low in the sky. It is in the east.
2. It is noon. The sun is high in the sky.
3. I predicted the sun would be high in the sky at noon. It would move to the west. My prediction matched what I observed.
4. I predict the sun will move the same way tomorrow but it will appear a little bit later than today.

▼ Drawings can help you remember details, such as the names of constellations.

Constellations

What are the names for the patterns of stars?

1
Orion

2
Scorpius

3
Big Dipper

▼ Use your notebook to make drawings and explain main ideas.

Stars Appear to Move

The sun, moon, and many stars appear to move across the sky in an arch-shaped pattern. The first star of the handle of the Big Dipper appears to move this way.

# Earth Science

## Space Systems: Patterns and Cycles

From Earth, you can see the moon appear to move across the sky.

# The Sun

Where have you seen **stars**? From Earth, you can see many stars in the night sky. The **sun** is a star. From Earth, the sun can only be seen in the daytime.

Like all stars, the sun gives off light and heat. You should never look right at the sun. Its bright light can harm your eyes.

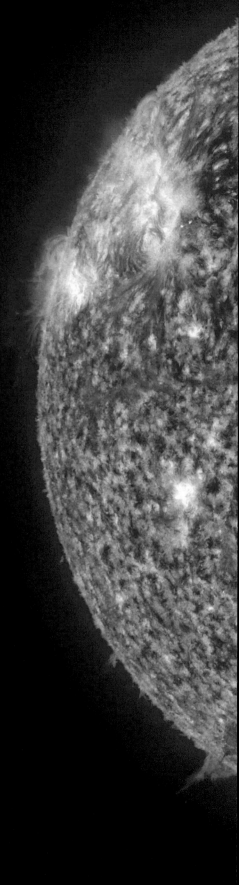

This photograph of the sun was taken with a special camera. The sun looks like a giant ball of fire.

**DCI** ESS1.A : The Universe and Its Stars. Patterns of the motion of the sun, moon, and stars in the sky can be observed, described, and predicted. (1-ESS1-1)
**CCC** Patterns. Patterns in the natural and human designed world can be observed, used to describe phenomena, and used as evidence. (1-ESS1-1)

## Wrap It Up! 📓 My Science Notebook

1. What is the sun?

2. Describe the sun. Like all stars, what does the sun give off?

# Day and Night

Daytime is light. Night is dark. Day and night happen again and again. They make a pattern.

Daylight comes from the sun. The sun appears to rise in the morning. The sky becomes light. The sun appears to move across the sky during the day. Then the sun appears to set in the evening. The sky gets dark.

The sun lights up Earth. The day sky is bright. You cannot see other stars in the daytime.

**DCI** ESS1.A: The Universe and Its Stars. Patterns of the motion of the sun, moon, and stars in the sky can be observed, described, and predicted. (1-ESS1-1)
**CCC** Patterns. Patterns in the natural and human designed world can be observed, used to describe phenomena, and used as evidence. (1-ESS1-1)

The night sky is dark. On a clear night you can see many stars, but not the sun.

## Wrap It Up! 📓 My Science Notebook

1. What is a pattern?

2. Describe how day and night make a pattern.

# The Sun in the Sky

In the morning the sun appears to be low toward the eastern part of the sky. In the middle of the day it appears higher in the sky. Late in the day the sun appears to be low in the sky again. You see it if you look toward the west.

**DCI** **ESS1.A: The Universe and Its Stars.** Patterns of the motion of the sun, moon, and stars in the sky can be observed, described, and predicted. (1-ESS1-1)

**CCC** **Patterns.** Patterns in the natural and human designed world can be observed, used to describe phenomena, and used as evidence. (1-ESS1-1)

In the morning the sun appears to be low in the sky.

The sun appears to move on an arch-shaped path. It is highest at noon.

The sun appears low in the sky again late in the day.

## Wrap It Up! My Science Notebook

1. Describe the pattern of the sun in the sky.

2. What can you predict about the sun tomorrow morning? What about the morning after that?

# Investigate

# The Sun

**?** **What can you observe about the position of the sun?**

The sun appears to move across the sky. You can observe this pattern. You can describe this pattern.

## Materials

crayons

paper

 **PE** 1-ESS1-1. Use observations of the sun, moon, and stars to describe patterns that can be predicted.

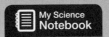
**1** Look at the sky in the morning. Where is the sun? Draw what you see.

**2** Look at the sky two hours later. Where is the sun? Draw what you see.

**3** Predict where you will see the sun in two more hours. Predict where you will see it in four hours, and then six hours. Draw your predictions.

**4** Look at the sky every two hours. Where is the sun? Draw what you see.

### Wrap It Up!  My Science Notebook

1. What did you observe about the sun in the morning and at noon?

2. Did your prediction match what you saw in the afternoon? Explain.

3. Use the pattern you saw to predict how the sun will move tomorrow.

# The Moon

You can often see the **moon** in the sky. Sometimes you can see the moon at night. Sometimes you can see the moon in the daytime, too.

The moon is not like the sun. It is not a star. It does not give off its own light. It reflects light from the sun.

You can see the moon on many days, too. You have to look harder to find it. It is not as bright as the sun.

**DCI** **ESS1.A: The Universe and Its Stars.** Patterns of the motion of the sun, moon, and stars in the sky can be observed, described, and predicted. (1-ESS1-1)

**CCC** **Patterns.** Patterns in the natural and human designed world can be observed, used to describe phenomena, and used as evidence. (1-ESS1-1)

You can see the moon on many nights. The moon is easy to see in the dark sky.

## Wrap It Up!

1. When can you see the moon?

2. When is the moon easier to see? Why?

# The Moon in the Sky

The moon is not always in the same place.
The moon first appears low in the sky.
You can see it first toward the east. Then
it appears high in the sky. Later it is lower
in the sky again. This time it is toward
the west.

**DCI ESS1.A: The Universe and Its Stars.** Patterns of the motion of the sun, moon, and stars in the sky can be observed, described, and predicted. (1-ESS1-1)
**CCC Patterns.** Patterns in the natural and human designed world can be observed, used to describe phenomena, and used as evidence. (1-ESS1-1)

The moon first appears low in the sky.

The moon appears to move on an arch-shaped path across the sky.

Later, the moon appears low in the sky again.

## Wrap It Up!

1. Describe the pattern of the moon moving in the sky.

2. What can you predict about the moon each time you can see it?

# The Moon

**?** **What can you observe about the position of the moon?**

The moon seems to move in the sky. You can observe the moon. You can describe its pattern of movement.

## Materials

crayons

paper

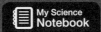
**1** Look at the sky. Where is the moon? Draw what you see.

**2** Look at the sky one hour later. Where is the moon? Draw what you see.

**3** Predict where the moon will be one, two, and three hours later. Draw your predictions.

**4** Go outside each hour for three hours. Find the moon in the sky. Draw what you see.

**Wrap It Up!**

1. Describe where you first saw the moon.

2. Use your pictures. Describe how the moon appeared to move.

3. Predict how the moon will appear to move tomorrow.

# Stories in Science

Katherine Johnson worked for the space program. Her math work was very important.

Katherine Johnson used math. She planned a path for a rocket to take around Earth.

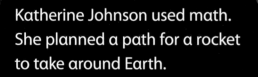

**DCI** ESS1.A: **The Universe and Its Stars.** Patterns of the motion of the sun, moon, and stars in the sky can be observed, described, and predicted. (1-ESS1-1)
**CCC** Patterns. Patterns in the natural and human designed world can be observed, used to describe phenomena, and used as evidence. (1-ESS1-1), (1-ESS1-2)
**NS** Scientific Knowledge Assumes an Order and Consistency in Natural Systems. Science assumes natural events happen today as they happened in the past. (1-ESS1-1)

# Katherine Johnson: Space Hero

Katherine Johnson was a math whiz kid. She liked to count. She counted steps. She counted dishes. She counted and counted. She went to high school at ten. She went to college at 14.

She worked for the space program. Katherine used math to help send the first American around Earth. She used math to help land men on the moon. She had to compare where the sun, moon, and Earth were. She had to know the path of the moon around Earth. Then she put her math to work.

## Wrap It Up!

1. Why do you think Katherine Johnson can be called a space hero?

2. How did Katherine use math for the space program?

# Stars

The sky is full of stars. You can see them on clear nights. They look like tiny points of light. Some are dim. Others are brighter.

Each star is like the sun. It gives off its own light and heat. The stars you see at night are very far away. That makes them look small.

**DCI** ESS1.A: The Universe and Its Stars. Patterns of the motion of the sun, moon, and stars in the sky can be observed, described, and predicted. (1-ESS1-1)
**CCC** Patterns. Patterns in the natural and human designed world can be observed, used to describe phenomena, and used as evidence. (1-ESS1-1)

Stars are always in the sky. You cannot see them during the day. The sun makes the sky too bright for the stars to be seen.

Stars shine all the time. You can only see them in the dark night sky.

## Wrap It Up!  My Science Notebook

1. When can you observe stars?

2. Why can't you see stars during the day?

# Star Patterns

People look for patterns in the stars. Some stars look like they are in groups. They look close together in the sky. People imagine that stars are connected together to make a pattern. Star patterns have names. The patterns help people remember where the stars are in the sky.

**North Star**

This star pattern is called the Big Dipper. It looks like a cup with a long handle. The Big Dipper points to the North Star.

**DCI** ESS1.A: The Universe and Its Stars. Patterns of the motion of the sun, moon, and stars in the sky can be observed, described, and predicted. (1-ESS1-1)
**CCC** Patterns. Patterns in the natural and human designed world can be observed, used to describe phenomena, and used as evidence. (1-ESS1-1)

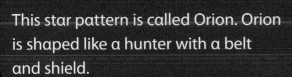

This star pattern is called Orion. Orion is shaped like a hunter with a belt and shield.

This star pattern is called Scorpius. Scorpius is shaped like a scorpion with a curled tail.

## Wrap It Up!

1. How can you use stars to make a pattern?

2. What are the names of some star patterns?

3. How do people use star patterns?

North Star

**DCI** ESS1.A: The Universe and Its Stars. Patterns of the motion of the sun, moon, and stars in the sky can be observed, described, and predicted. (1-ESS1-1)
**CCC** Patterns. Patterns in the natural and human designed world can be observed, used to describe phenomena, and used as evidence. (1-ESS1-1)

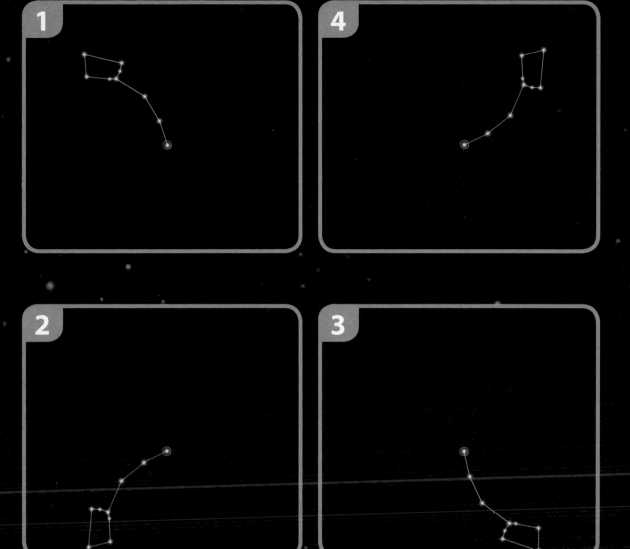

Polaris is like the center of a wheel. The Little Dipper appears to spin around it. The drawings show how the Little Dipper appears to make this circle every 24 hours. People cannot see the whole circle because stars are not visible during the day.

## Wrap It Up!

1. Describe the Little Dipper.

2. Where is the North Star in the Little Dipper?

# Patterns of Motion

The sun and moon appear to move across the sky in an arch-shaped pattern. Most stars seem to move in a similar pattern.

The star at the end of the Big Dipper's handle is called Alkaid. On nights when you can see the Big Dipper, you can observe Alkaid. It appears to move on an arch-shaped path across the sky.

People can use special cameras to take photos of the starry sky.

DCI ESS1.A: The Universe and Its Stars. Patterns of the motion of the sun, moon, and stars in the sky can be observed, described, and predicted. (1-ESS1-1)
CCC Patterns. Patterns in the natural and human designed world can be observed, used to describe phenomena, and used as evidence. (1-ESS1-1)

Alkaid first appears low in the sky.

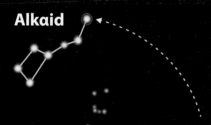

Alkaid appears to move across the sky on an arch-shaped path.

Later, Alkaid appears low in the sky.

## Wrap It Up!

1. Where is Alkaid in the Big Dipper?

2. Describe Alkaid's pattern of motion.

121

# The Night Sky

**?** **How do some stars seem to move in the night sky?**

Stars appear to move in the night sky. You can predict where you will see star patterns in the sky.

You will look at a star pattern called Cepheus (sē-fē-us).

## Materials

**construction paper**

**night sky model**

PE 1-ESS1-1. Use observations of the sun, moon, and stars to describe patterns that can be predicted.

**1** Use your night sky model. Turn the small circle. Point the arrow at *A*.

**2** Find Cepheus. Draw it in your notebook.

**3** Point the arrow at *B*. Draw Cepheus again.

**4** Predict how you think Cepheus will look when the arrow points to *C*. Point the arrow at *C*. Draw what you see.

## Wrap It Up!

1. Do your observations support your prediction? Tell why or why not.

2. Describe how Cepheus appears to move.

3. Look again at the model. Turn it again. Describe what you observe about the position of the North Star.

STEM

SCIENCE
TECHNOLOGY
ENGINEERING
MATH

ENGINEERING PROJECT

# Design a Sundial

How can you tell time without a clock? You can use shadows. You will make a simple sundial. Then design your own sundial. A sundial is simple. It has a pole that makes a shadow.

Earth spins. The sun appears to move in the sky. Shadows move. They change size. You can mark where the shadow is every hour. Then you can tell the time.

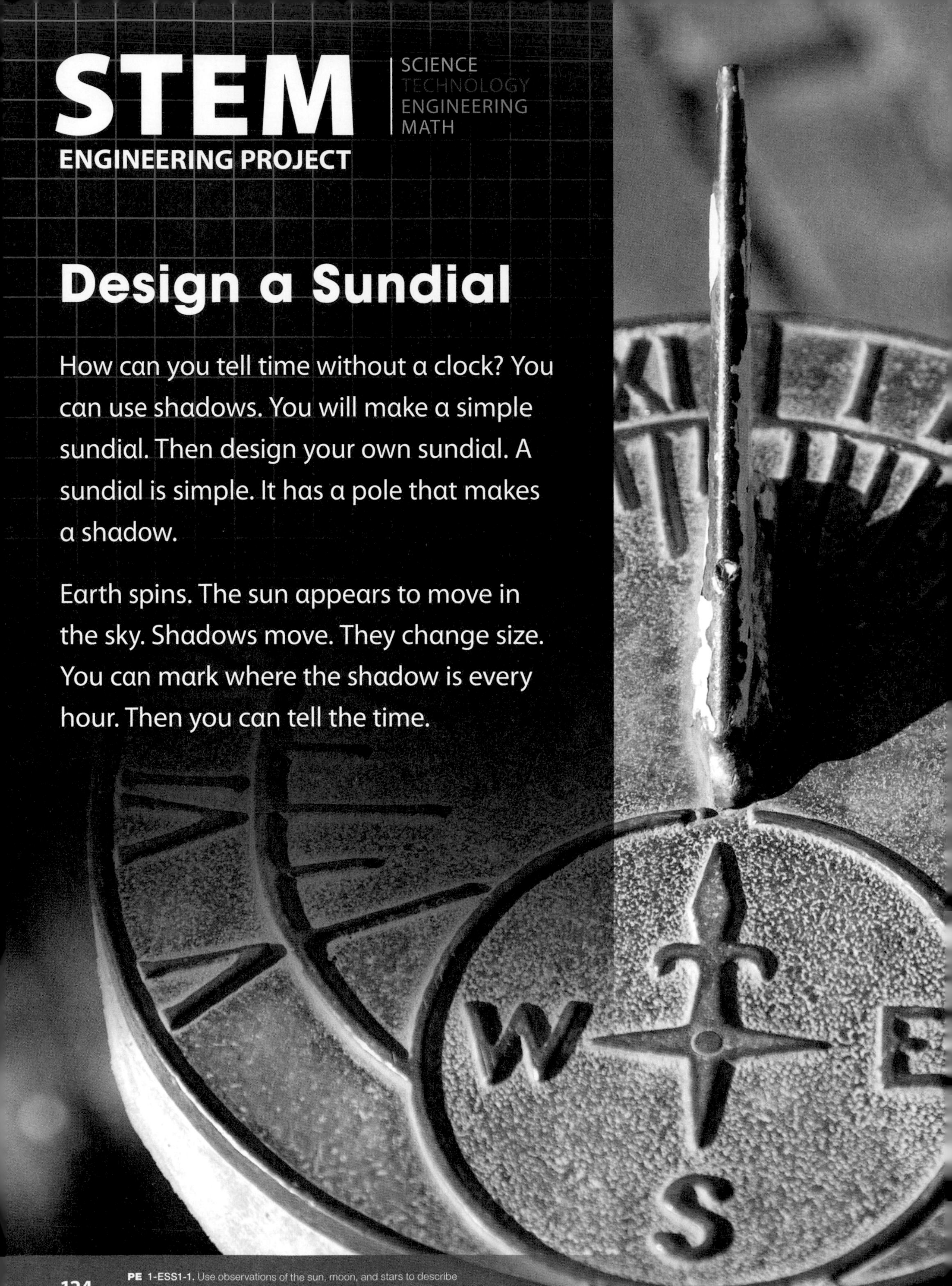

**PE** 1-ESS1-1. Use observations of the sun, moon, and stars to describe patterns that can be predicted.

Earth spins. The sun appears to move across the sky. The shadow on the sundial helps you tell time.

# The Challenge

Design your own sundial. Use it where there is no clock. You could use it outdoors.

**1** **Define the problem.**

What problem can you solve with a sundial? Write the problem in your science notebook. Describe how your sundial will solve the problem. Your teacher will show you the materials you may use. Your teacher will tell you how much time you have.

**2** **Design a solution.**

Choose materials. Draw your sundial design. Show your team. Discuss each design. Choose one design you think will be best. Draw your final design.

**3** **Test your solution.**

Build your sundial. Then test it. Does it work? If not, make changes. Draw your new design. Make changes. Then test it again.

**4** **Refine or change your solution.**

Talk with your team. Can you make your sundial better? Write your ideas in your science notebook. Show your sundial to the class. Tell how it worked.

# Seasons

Weather changes through the year. In many places the temperature can become colder for months at a time. Then it becomes warmer.

In some places there are four **seasons.** They are winter, spring, summer, and fall. The seasons follow a pattern. They come in the same order, year after year.

Many trees like this maple change as the seasons change.

**DCI** ESS1.B: Earth and the Solar System. Seasonal patterns of sunrise and sunset can be observed, described, and predicted. (1-ESS1-2)
**CCC** Patterns. Patterns in the natural and human designed world can be observed, used to describe phenomena, and used as evidence. (1-ESS1-2)

Winter is the coolest season.

The temperature gets cooler in the fall.

The temperature gets warmer in the spring.

Summer is the warmest season.

## Wrap It Up! My Science Notebook

1. Describe the pattern of the seasons.

2. How many winter seasons happen in three years?

# Light and the Seasons

As the seasons change, the amounts of daylight and darkness change, too. **Sunrise** happens earliest and **sunset** happens latest in summer. Summer has the most hours of daylight. Fall has fewer hours of daylight than summer. In winter, the sun rises late and sets early. Winter has the fewest hours of daylight. In spring, the number of hours of daylight increase again.

It is 6:00 p.m. on a summer evening. It is still light.

**DCI** ESS1.B: Earth and the Solar System. Seasonal patterns of sunrise and sunset can be observed, described, and predicted. (1-ESS1-2)

**CCC** Patterns. Patterns in the natural and human designed world can be observed, used to describe phenomena, and used as evidence. (1-ESS1-2)

It is 6:00 p.m. on a winter evening. It is already dark.

## Wrap It Up! 📔 My Science Notebook

1. Which season has the most hours of daylight? Which has the least?

2. Which two seasons can have days with about the same number of daylight hours?

3. You wake at sunrise on a spring morning. What can you predict about the sunrise the next day?

# Citizen Science

Flowers bloom. They form fruits and seeds. This happens at different times for different plants.

It is fall. Leaves of different trees can change color at different times.

These students are observing plants. They collect and record data. They share data with scientists.

**DCI** ESS1.B: Earth and the Solar System. Seasonal patterns of sunrise and sunset can be observed, described, and predicted. (1-ESS1-2)
**SEP** Planning and Carrying Out Investigations. Make observations (firsthand or from media) to collect data that can be used to make comparisons. (1-ESS1-2)

**CCC** Patterns. Patterns in the natural and human designed world can be observed, used to describe phenomena, and used as evidence. (1-ESS1-2)

# Project BudBurst

## What Is Citizen Science?

Temperatures on Earth are rising. How does this affect plants? Scientists want to know. They want your help! People across the country can collect data for scientists. This is called citizen science.

In Project BudBurst, people observe plants. They record changes through the seasons. They share data with scientists. Scientists look for patterns. They find out how temperature change affects plants.

You might observe when leaves come out in spring or drop in fall. You might observe when flowers bloom. Your teacher will give you the information you need.

## Wrap It Up!  My Science Notebook

1. What kind of plant or plants did you observe? In what season or seasons did you make observations?

2. What data did you collect? What was the most interesting thing you found?

# Make Observations

You can make observations to see how the time of sunrise and sunset changes during the year.

**1  Ask a question.** `My Science Notebook`

Sheena was in first grade. She got up one morning at her usual time. She saw that it was just getting light. In the summer it was light when she got up. Later that day, it got dark earlier than it did in the summer. Could the time of sunrise and sunset change?

**2  Conduct an investigation.**

- Think about what you have learned. How might the time of sunrise and sunset change? Write or draw your ideas.

- Think about how you can collect data about the number of daylight hours throughout the year. How can you record your data?

- Make a plan. Carry out your plan.

## 3 Analyze your results.

Look at your data. Look for patterns.
What do they show?

## 4 Explain your results.

Share your data with a partner.
Explain what your data show.
Which times of year had the most
daylight? Which had the least?
Tell how you know.

# Astronomer

Knicole Colón knew when she was 12 years old that she wanted to study astronomy. An astronomer is a scientist. Astronomers study objects in space. They study the moon and sun. They also study stars and planets.

Knicole collects information about stars and planets. She uses tools such as telescopes. Telescopes help her make observations. Knicole thinks there may be other planets like Earth. She hopes that her work will help discover them.

**NS** Scientific Investigations Use a Variety of Methods. Science investigations begin with a question. (1-PS4-1).
**NS** Scientific Investigations Use a Variety of Methods. Scientists use different ways to study the world. (1-PS4-1)
**NS** Scientific Knowledge Is Based on Empirical Evidence. Scientists look for patterns and order when making observations about the world. (1-LS1-2)

NATIONAL GEOGRAPHIC | Explorer

**Knicole Colón** is an astronomer. She wants to discover new planets that may support living things.

Knicole loves visiting different telescopes. Some are located on tall mountains and in other beautiful places.

# Check In

Way to go! You have completed *Earth Science.* Think about what you learned. Here is a checklist to help you check your progress. Look through your science notebook to find examples of the items. Tell how well you did. Write in your science notebook.

▼ Read the list. Think about how you used your science notebook.

**For each item, select the choice that is true for you:  A.** Yes  **B.** Not Yet

- I made drawings of new science words and main ideas.

- I labeled drawings. I wrote to explain ideas.

- I collected photos, news stories, and other objects.

- I used tables, charts, or graphs to record observations.

- I recorded reasons for explanations and conclusions.

- I wrote about what scientists and engineers do.

- I asked new questions.

- I did something else. (Tell about it.)

## Reflect on Your Learning

1. What did you learn that changed the way you think about something?

2. Is there anything you don't fully understand? What can you do to help yourself understand better?

Nalini uses ropes and cables to climb a tree. She climbs hundreds of feet to the forest canopy.

**Nalini Nadkarni**  Forest Ecologist
National Geographic Explorer

# Let's Explore!

Scientists investigate. They find answers to questions. I study a group of plants called orchids. They grow high in the trees. They do not have roots in the ground. But they get water. They get light. I observe them and look for patterns. Then I explain how the plants survive. As you read, act like a scientist. Look for patterns to help make explanations.

Physical science includes the study of sound and light. I study how plants get light from the sun. They need light to survive. Here are some questions to answer in *Physical Science*:

- What causes a guitar to make a sound?

- How can you use your voice to make a balloon vibrate?

- What happens when light shines on different objects?

Be sure to ask new questions. Look for answers. Let's check in again to review what you have learned!

## Sounds

| Object | Makes sound when plucked |
|---|---|
| string | yes |
| rubber band | yes |
| rope | no |
| shoelace | no |

## Reflect Light

Light bounces off a mirror to your eye. You see the flashlight and the light shining in the mirror.

▶ Write about patterns in different areas of science.

### Earth Science
### Patterns in the Sky

The sun, moon, and many stars appear to move in the sky. This is because Earth is spinning.

### Physical Science
### Patterns in Sound

Plucking a string makes a sound.
Plucking a rubber band makes a sound.
Plucking can make vibrations.
Vibrations make sound.

# Physical Science

## Waves: Light and Sound

Two musicians play instruments in Amer, a town in India.

# Vibrate and Make Sound

Pluck a guitar string. What happens? The string moves back and forth. When things move back and forth quickly, they **vibrate.** Things that vibrate make **sound.**

When the string vibrates, the air near the string vibrates, too. The vibrations move to your ears. Then you hear the sound.

**DCI** PS4.A: Wave Properties. Sound can make matter vibrate, and vibrating matter can make sound. (1-PS4-1)

**CCC** Cause and Effect. Simple tests can be designed to gather evidence to support or refute student ideas about causes. (1-PS4-1)

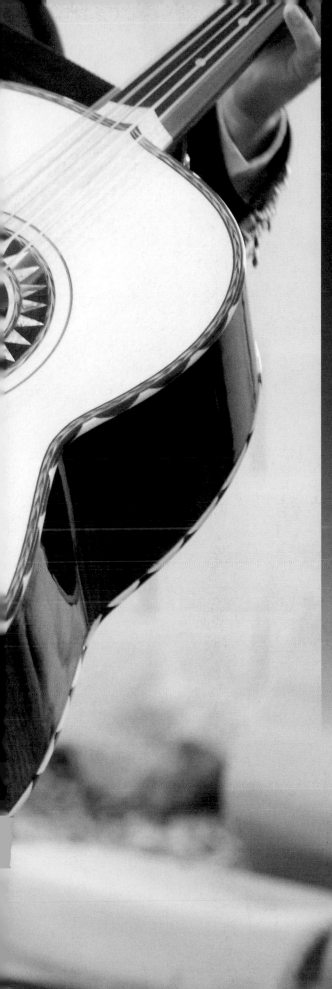

## Feel Vibrations

**1** Touch your throat as you talk. Listen to the sound.

**2** Repeat. This time, talk louder.

**?** **What do you feel with your fingers? How does it change when you talk louder?**

## Wrap It Up!

1. What does vibrate mean?

2. List some other things that vibrate. What sounds do they make?

# Sound

**?** **How can vibrations make sound?**

A rubber band can stretch. You can pluck it. It will vibrate. You can observe how a rubber band vibrates.

## Materials

**2 rubber bands**

**cardboard box**                    **hand lens**

**DCI** PS4.A: Wave Properties. Sound can make matter vibrate, and vibrating matter can make sound. (1-PS4-1)
**CCC** Cause and Effect. Simple tests can be designed to gather evidence to support or refute student ideas about causes. (1-PS4-1)

144

**1** Put on safety goggles. Choose one rubber band.

**2** Stretch the rubber band around the box.

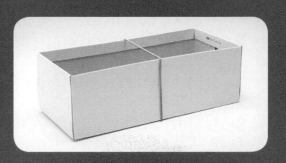

**3** Predict what will happen when you pluck the rubber band.

**4** Pluck the rubber band. Use the hand lens to see how it vibrates. Record what you see and hear. Try it again with the other rubber band.

## Explore on Your Own

What would happen if you plucked two thick rubber bands of different sizes? Make a plan to investigate. Carry out your plan. Record your observations. Compare the results of your investigations.

## Wrap It Up!

1. What happened when you plucked the rubber bands? What did you see? What did you hear?

2. Compare the two rubber bands. How did they act differently?

# Plan and Investigate

You investigated rubber bands. You found that a rubber band can vibrate. Then it makes sound. Now it is your turn. How can you show that vibrating materials make sound?

**1** **Plan an investigation.**

Think of things that make sound when they vibrate. Think of how you can make them vibrate. Write down your ideas.

List some things to test. Make a plan. Draw a picture of what your test will look like.

Gather your materials.

**2** **Conduct an investigation.**

Carry out your plan. Observe what happens. Tell what you see and hear before and during your test. Record your data in a chart.

**PE** 1-PS4-1. Plan and conduct investigations to provide evidence that vibrating materials can make sound and that sound can make materials vibrate.

**3** **Analyze results.**

Look at your data. What are your results? Did the materials make sound when they were not vibrating? Did vibrating materials make sound? Tell what you observed that supports your answers.

**4** **Share your results.**

Show others how you made the materials vibrate. Explain what happened when the materials vibrated.

# STEM
ENGINEERING PROJECT

# Design a Drum

Vibrating materials can make a sound. Sound can make materials vibrate. A toy company wants to make a new game. The game uses a drum. You strike the drum. Objects on a plastic surface jump.

The drums make sound. It can make materials vibrate.

**PE 1-PS4-1.** Plan and conduct investigations to provide evidence that vibrating materials can make sound and that sound can make materials vibrate.
**PE K–2-ETS1-1.** Ask questions, make observations, and gather information about a situation people want to change to define a simple problem that can be solved through the development of a new or improved object or tool.

# The Challenge

Design and build a model of a toy drum. You will strike the drum. The sound must make plastic wrap vibrate. Two different objects on the plastic wrap must jump.

## 1 Define the problem.

Your teacher will give you materials. Write the problem you need to solve in your science notebook.

## 2 Design a solution.

Choose materials. Think about how you will strike the drum. Draw your design. Show your team. Discuss each design. Choose one design you think will be best. Draw your final design. Build your model.

## 3 Test your solution.

Test your model. Did it make two different objects on the plastic wrap jump? If not, change it. Draw your new design. Explain your changes.

## 4 Refine or change your solution.

Talk with your team. Can you make your model better? Write your ideas in your science notebook. Show your drum to the class. Tell how it worked.

# Sound Makes Things Vibrate

Things that vibrate make sound. Sound can make things vibrate, too. Drummers banging on these big drums make a loud sound. If you could touch the walls and floor in the room, you would probably feel the vibration from the sound.

**DCI PS4.A: Wave Properties.** Sound can make matter vibrate, and vibrating matter can make sound. (1-PS4-1)
**CCC Cause and Effect.** Simple tests can be designed to gather evidence to support or refute student ideas about causes. (1-PS4-1)

## Wrap It Up! 📓 My Science Notebook

1. What could you touch near a loud drum to feel the vibration from the sound?

2. What sounds might cause things to vibrate in your home?

# Vibration

**?** **How can you use sound to make an object vibrate?**

You have learned that sound can make objects vibrate. You can observe how the sound of your voice makes a balloon vibrate.

## Materials

**inflated balloon**          **paper towel tube**

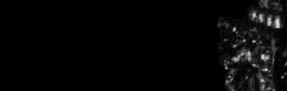

**DCI** PS4.A: Wave Properties. Sound can make matter vibrate, and vibrating matter can make sound. (1-PS4-1)
**CCC** Cause and Effect. Simple tests can be designed to gather evidence to support or refute student ideas about causes. (1-PS4-1)

152

**1** Work with a partner. Hold the balloon gently with your fingertips.

**2** While your partner talks quietly into one end of the tube, hold the balloon very close to the other end.

**3** Observe what you hear. Observe what you feel through the balloon. Record what you hear and feel.

**4** Switch places and repeat while your partner observes and records.

## Wrap It Up! My Science Notebook

1. What did you feel when you held the balloon?

2. What was the cause of the vibrations in the balloon?

# Plan and Investigate

You used sound to make a balloon vibrate. Now you will investigate vibrations with other materials. How can you show that sound makes materials vibrate?

**1** **Plan an investigation.** My Science Notebook

Think of things that make sound. Think of how you can use that sound to make something else vibrate. Write down your ideas.

Make a plan. Write down your steps. Draw a picture of what your test will look like.

Gather your materials.

**PE** 1-PS4-1. Plan and conduct investigations to provide evidence that vibrating materials can make sound and that sound can make materials vibrate.

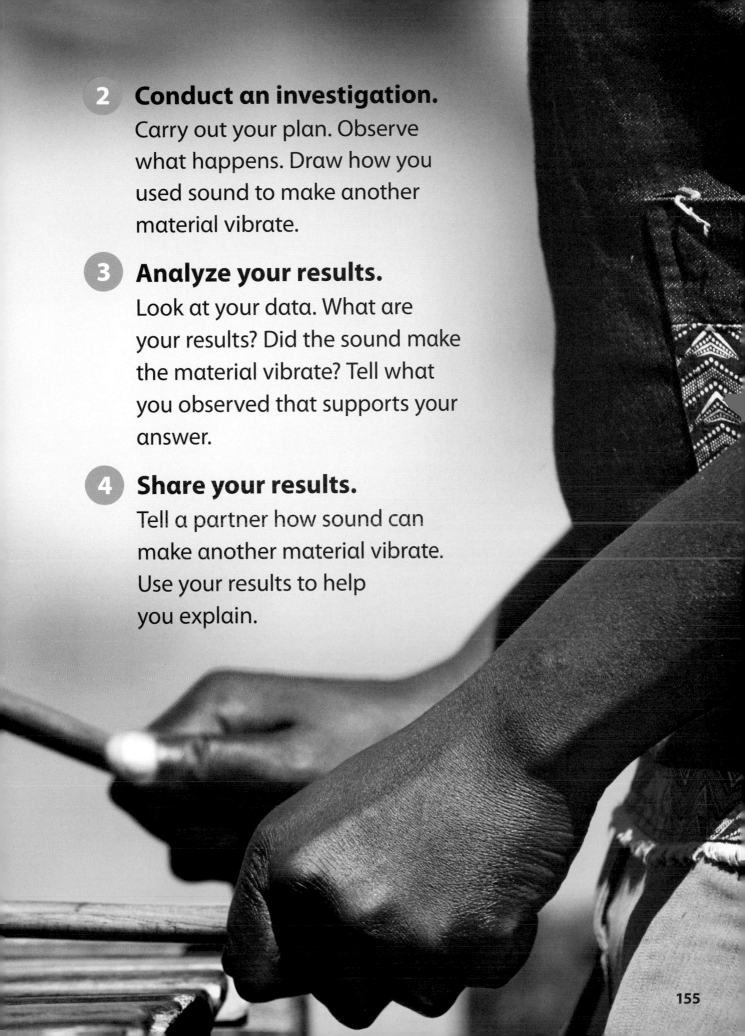

**2** **Conduct an investigation.**
Carry out your plan. Observe what happens. Draw how you used sound to make another material vibrate.

**3** **Analyze your results.**
Look at your data. What are your results? Did the sound make the material vibrate? Tell what you observed that supports your answer.

**4** **Share your results.**
Tell a partner how sound can make another material vibrate. Use your results to help you explain.

# Light

What do you see in the picture? There are deer, trees, and a grassy hill. You also see sunlight. **Light** makes it possible to see objects.

The sun gives off its own light. Light from the sun allows us to see during the day. At night it is dark. People cannot see when it is dark.

**DCI PS4.B: Electromagnetic Radiation.** Objects can be seen if light is available to illuminate them or if they give off their own light. (1-PS4-2)
**CCC Cause and Effect.** Simple tests can be designed to gather evidence to support or refute student ideas about causes. (1-PS4-2)

## Wrap It Up! 📓 My Science Notebook

1. What helps you see the objects in the picture on this page?

2. What might you see if this picture were taken on a very dark night?

# Light to See

Much of this cave is dark. People can only see where there is light. The diver uses a flashlight to make light. The diver can now see the cave.

**DCI** PS4.B: **Electromagnetic Radiation.** Objects can be seen if light is available to illuminate them or if they give off their own light. (1-PS4-2)

**CCC Cause and Effect.** Simple tests can be designed to gather evidence to support or refute student ideas about causes. (1-PS4-2)

A little light comes into the cave through its opening. Away from the opening, the cave is dark.

## Wrap It Up! 📓 My Science Notebook

1. Why can you see objects in this cave?

2. What could you see in the picture if the diver's flashlight was turned off?

159

# Investigate

# Light and Dark

**?** **Do you need light to see?**

You enter a dark room. What can you see? Turn on a light. What can you see then? You can investigate when you can see and when you cannot.

## Materials

| **cardboard box with two holes** | **flashlight** | **masking tape** |
| --- | --- | --- |
|  |  |  |

 **PE 1-PS4-2.** Make observations to construct an evidence-based account that objects in darkness can be seen only when illuminated.

**1** Find the big hole on the top of the box. Put the flashlight over it. Use tape to attach the flashlight.

**2** Cover your eyes. Your partner will find a mystery object in the room and put it in the box. Do not peek!

**3** Leave the flashlight off. Look through the hole. Draw what you see. Turn on the flashlight and look again. Draw what you see.

**4** Switch with your partner. Repeat steps 2 and 3.

## Wrap It Up!

1. Tell how the object appears when it is in the dark.

2. Tell how the object appears when it is in a space with light.

3. Explain how the light changed what you saw.

# Shining Through

Look at the sea turtle. You can see it even though it is underwater. The shallow water is clear. **Clear** materials do not block any light. They let light pass through.

Light passes through the water, so you can see the sea turtle and the corals beyond it.

**DCI** PS4.B: Electromagnetic Radiation. Some materials allow light to pass through them, others allow only some light through and others block all the light and create a dark shadow on any surface beyond them, where the light cannot reach. Mirrors can be used to redirect a light beam. (1-PS4-3)
**CCC Cause and Effect.** Simple tests can be designed to gather evidence to support or refute student ideas about causes. (1-PS4-3)

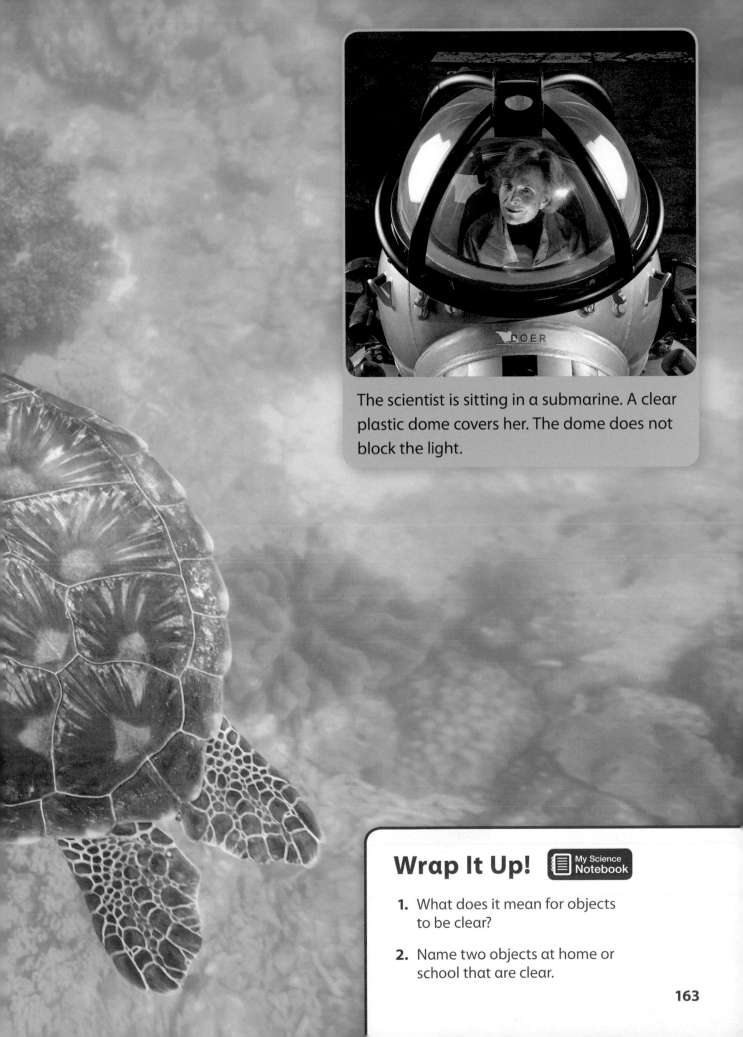

The scientist is sitting in a submarine. A clear plastic dome covers her. The dome does not block the light.

## Wrap It Up! 📓 My Science Notebook

1. What does it mean for objects to be clear?

2. Name two objects at home or school that are clear.

163

# Blocking Some Light

Some materials block some light. They still let some light through. This butterfly's wings let some light pass through. You can see the flower's petals through the butterfly's wings.

**DCI** **PS4.B: Electromagnetic Radiation.** Some materials allow light to pass through them, others allow only some light through and others block all the light and create a dark shadow on any surface beyond them, where the light cannot reach. Mirrors can be used to redirect a light beam. (1-PS4-3)

**CCC** **Cause and Effect.** Simple tests can be designed to gather evidence to support or refute student ideas about causes. (1-PS4-3)

## Wrap It Up! 📓 My Science Notebook

1. How do you know the butterfly's wings let some light pass through them?

2. Name three things at home or school that block some light.

# Blocking All Light

Some objects block all light. No light passes through them. Things that block light make a shadow.

Can you see the dragonfly's shadow in the picture? A **shadow** is a dark place beside or under an object. Light is on one side of an object. The shadow is on the opposite side. The shadow's shape changes when the object moves or the light moves.

**DCI  PS4.B: Electromagnetic Radiation.** Some materials allow light to pass through them, others allow only some light through and others block all the light and create a dark shadow on any surface beyond them, where the light cannot reach. Mirrors can be used to redirect a light beam. (1-PS4-3)
**CCC  Cause and Effect.** Simple tests can be designed to gather evidence to support or refute student ideas about causes. (1-PS4-3)

Light passes through the dragonfly's wings but not through its body.

## Making Shadows

**1** Have a partner stand behind you with a flashlight. Face the wall. Move your hands in front of the light. See what happens.

**2** Repeat with the flashlight turned off. See what is different.

**?** **Where did a shadow appear? What happened when the flashlight was turned off?**

## Wrap It Up!  My Science Notebook

1. What is a shadow?

2. How does your body make a shadow?

3. If you stand facing the sun, where will your shadow be?

167

# Reflecting Light

Some objects **reflect** light. They make light bounce back. Smooth, shiny objects reflect light clearly. That is why you can see your reflection in a mirror.

Look at the cheetahs in the picture. You can see their images reflected in the water. The water is smooth like a mirror. Light from the sun bounces back from the surface of the water.

**DCI PS4.B: Electromagnetic Radiation.** Some materials allow light to pass through them, others allow only some light through and others block all the light and create a dark shadow on any surface beyond them, where the light cannot reach. Mirrors can be used to redirect a light beam. (1-PS4-3)
**CCC Cause and Effect.** Simple tests can be designed to gather evidence to support or refute student ideas about causes. (1-PS4-3)

## Reflections

**1** Shine a flashlight on a mirror. See what happens. Now move the light. See what happens.

**2** Now move the mirror. Shine the flashlight on it again.

**?** **What happened when you moved the light? The mirror?**

## Wrap It Up!  My Science Notebook

1. What happens when light shines on a mirror?

2. How can water act like a mirror?

# Plan and Investigate

You have learned that light does different things when it shines on different materials. Now it is your turn. How can you show what happens to light when it shines on different objects?

**1** **Plan an investigation.** 📓 My Science Notebook

Think about what light does when it shines on different objects.

List some things to test. Make a plan. Draw a picture of what your test will look like.

Gather your materials.

**2** **Conduct an investigation.**

Carry out your plan. Draw what you do during the steps of your investigation. Draw what you observe for each material you test.

**PE** 1-PS4-3. Plan and conduct investigations to determine the effect of placing objects made with different materials in the path of a beam of light.

### 3 Analyze results.

Look at your data. What are your results? How did the materials and light interact? Did light reflect off some materials? Which ones? Tell what you observed that supports your answers.

### 4 Share results.

Show your drawing to others. Tell how you did your tests. Explain what light did in each test.

Look at a classmate's results. Did you do the same kinds of tests? Did you observe the same things? What can you learn from what your classmate did?

# People Communicate

When you send or receive information, you **communicate.** People communicate by talking and writing. They use gestures like nodding or smiling.

People use devices to communicate over long distances. They can use computers to send email. These boys are using their cell phones to talk and send text messages or pictures.

**DCI** PS4.C: Information Technologies and Instrumentation. People also use a variety of devices to communicate (send and receive information) over long distances. (1-PS4-4)
**CETS** Influence of Engineering, Technology, and Science on Society and the Natural World. People depend on various technologies in their lives; human life would be very different without technology. (1-PS4-4)

# SCIENCE in a SNAP

## Use Flashlights to Communicate

yes                    no

**1** Stand across the room from a partner. Ask three questions. Have your partner use the yes/no code to answer.

**2** Trade roles, and repeat step 1.

**?** Did you understand your partner's answers correctly? Did your partner understand yours?

## Wrap It Up!

1. You and a friend are in the same room. How can you communicate?

2. A relative lives far away. How can you stay in touch?

Aydogan Ozcan is an electrical engineer. He is also a National Geographic Explorer. He invented a microscope. It can attach to a cell phone.

**DCI** PS4.C: Information Technologies and Instrumentation.
People also use a variety of devices to communicate (send and receive information) over long distances. (1-PS4-4)

**DCI** ETS1.A: Defining and Delimiting Engineering Problems.
A situation that people want to change or create can be approached as a problem to be solved through engineering. (K–2-ETS1-1)

**CETS** Influence of Engineering, Technology, and Science on Society and the Natural World. People depend on various technologies in their lives; human life would be very different without technology. (1-PS4-4)

# Using Cell Phones in New Ways

Many people live far from doctors. They cannot get help when they are sick.

Aydogan Ozcan is changing that. He invented a new tool. You put it on a cell phone. You use the phone as a microscope. You can look at blood samples. Doctors who are far away can look at the samples, too. They can find out what is causing sickness. They can choose the right medicine. The tool will save time and money. It will help save lives.

## Wrap It Up!

1. How does Aydogan Ozcan's invention change how a cell phone can be used?

2. How will Ozcan's invention help people?

## Investigate

# Communicating with Sound

**? How can you communicate with sound?**

People use telephones to communicate. In some phones, sound travels through wires from one phone to another. You can model how this kind of communication works.

## Materials

**two cups with slits**          **string and paper clips**

**DCI** PS4.C: Information Technologies and Instrumentation. People also use a variety of devices to communicate (send and receive information) over long distances. (1-PS4-4)
**CETS** Influence of Engineering, Technology, and Science on Society and the Natural World. People depend on various technologies in their lives; human life would be very different without technology. (1-PS4-4)

**1** Put a paper clip through the slit in each cup.

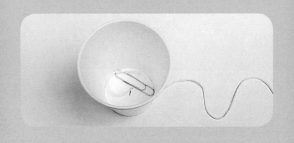

**2** Work with a partner. Hold the two cups apart. Do not pull too hard.

**3** Talk softly into your cup. Have your partner listen into the other cup. Then listen while your partner talks softly.

**4** Record the messages in your notebooks. Compare with your partner to find out if you recorded them correctly.

## Wrap It Up!

1. Did you receive your partner's message correctly? Did your partner receive your message correctly?

2. Use what you know about sound. Tell how the sound got from your cup to your partner's cup.

177

# Design a Device

People communicate in many ways. They talk. They nod. They use telephones to talk and send photos to each other.

Now it is your turn. How can you use sound or light to communicate with a partner?

**1 Design your device.**  My Science Notebook

Work with a partner. How can you send a message across the room? What materials will you need?

Write down your ideas. Draw your device. Label what each part will do.

Gather your materials. Build your device.

**PE 1-PS4-4.** Use tools and materials to design and build a device that uses light or sound to solve the problem of communicating over a distance.

**2** **Test your device.**

Use your device to send a message. Take turns. Record the message you send. Record the message you receive.

**3** **Refine or change your device.**

Study your results. Did your partner receive your message? Did you receive your partner's? Can you make your device better? Record your ideas. Try it!

**4** **Share your results.**

Show your class how your device works. Explain all the parts. Explain how people can use sound and light to communicate.

# Photographer

Gabby Salazar is a photographer. A photographer takes pictures. It is important for photographers to know about light. They use light in their pictures. The way light shines can make a picture look good or bad.

Gabby travels around the world taking pictures of nature. She spent ten months in Peru. People there are working to protect a rain forest. She took pictures of the plants, animals, and people that live there. Protecting wildlife is important to Gabby.

**Gabby Salazar** is a nature photographer. She works to help protect plants and animals. Gabby also enjoys teaching photography to children.

# Check In  My Science Notebook

Way to go! You have completed *Physical Science.* Think about what you learned. Here is a checklist to help you check your progress. Look through your science notebook to find examples of the items. Tell how well you did. Write in your science notebook.

▼ Read the list. Think about how you used your science notebook.

**For each item, select the choice that is true for you: A. Yes B. Not Yet**

- I made drawings of new science words and main ideas.

- I labeled drawings. I wrote to explain ideas.

- I collected photos, news stories, and other objects.

- I used tables, charts, or graphs to record observations.

- I recorded reasons for explanations and conclusions.

- I wrote about what scientists and engineers do.

- I asked new questions.

- I did something else. (Tell about it.)

## Reflect on Your Learning  My Science Notebook

1. Choose one science idea that you would like to learn more about. What questions can you ask?

2. Think about the science ideas that you learned. Which do you think is the most important and why?

# More to Explore

I still love climbing trees! I share my love of nature with other people. As a scientist, I ask questions and do investigations. There are always more questions to ask!

Look back in your science notebook. What were some interesting things you learned? Were you surprised by anything you learned? What important things did you learn? Share your thoughts with classmates. Think about what you want to learn more about. Then ask questions. Look for answers. Keep *Exploring Science.* Always be curious!

Many scientists work in teams. Nalini works with other scientists to study rain forests.

# Science Safety

Science safety is important. Follow all science safety rules to stay safe. Always ask an adult for help. Report any accidents right away.

## Science Safety Rules

- Follow all directions for lab procedures.
- Keep your area neat.
- Keep hands away from your eyes and mouth.
- Do not eat or drink anything in the science room.
- Tell your teacher about any allergies you have.
- Feet should be covered; no sandals.
- Wear goggles or gloves when told.
- Tie back loose or long hair.

## In the Lab Space

- Where is the first aid kit?
- Where is the fire blanket?
- What do you do if your clothes catch fire? (Stop-Drop-Roll)
- Are you wearing lab clothing or gear?

## During the Lab

- Ask your teacher for help, if needed.
- Handle science materials carefully.
- Be responsible with living things.
- Take care of plants and animals.

## During Cleanup

- Close all containers.
- Return materials to their correct locations.
- Throw out used gloves.
- Wash hands with soap and water.
- Ask an adult to place broken glass in a sealed container.

# Tables and Graphs

Scientists and engineers record different kinds of information. They record measurements and facts about their observations. This is called data. Scientists and engineers need to organize their data, so they use tables and graphs. This helps them share the data. As you do investigations, you'll collect data. You can share your data in a table or graph.

## Tables

A table is a set of **rows** and **columns.** Rows and columns set up a simple grid. Every piece of information has a place.

| Properties of Classroom Objects | | | |
| --- | --- | --- | --- |
| **Hard** | **Soft** | **Smooth** | **Rough** |
| desk | flag | laptop cover | sandstone rock |
| chair | cushion | magnifying lens | pinecone |
| book | fabric letters | white board | doormat |
| tablet | sponges | window glass | sandpaper |

Each property has its own column.

Every item has its own cell or place in the table.

You can add new information to a table by adding more rows or columns. What information does this table help organize?

# Bar Graphs

A bar graph is formed with two lines, the **x-axis** and **y-axis.** One line extends up. One line extends to the right. Bars sit on the line that goes across. You can compare the bars in a bar graph. Look at the height of each bar to compare the values, or numbers.

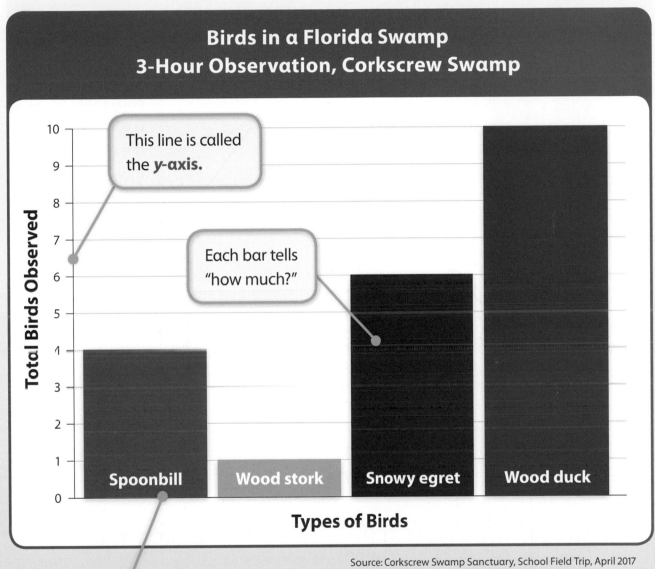

**Birds in a Florida Swamp**
**3-Hour Observation, Corkscrew Swamp**

This line is called the **y-axis.**

Each bar tells "how much?"

Total Birds Observed

Spoonbill   Wood stork   Snowy egret   Wood duck

**Types of Birds**

Source: Corkscrew Swamp Sanctuary, School Field Trip, April 2017

This line is called the **x-axis.**

What does each bar tell about in this graph? Which bar has the highest value? Which has the least value?

# Line Graphs

Like a bar graph, a line graph has an x-axis and y-axis, which form an *L* shape. Instead of a bar, each piece of data gets a dot, or point.

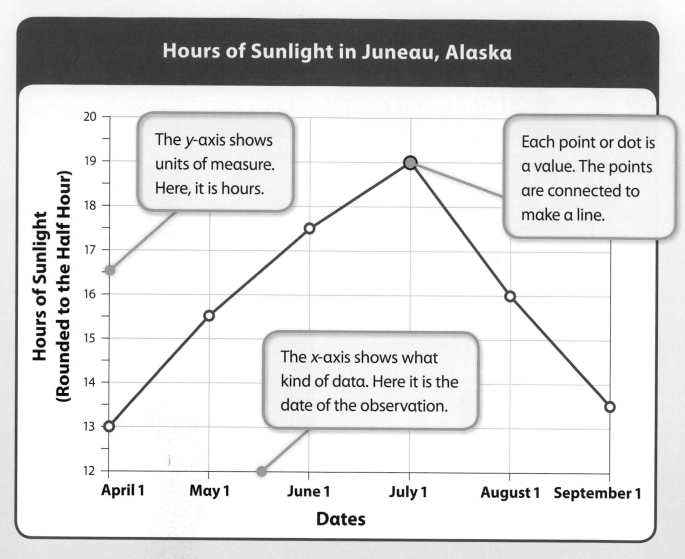

A line graph can be used to show changes over time. This line graph shows the number of hours of sunlight across six months. Which month had the most hours of sunlight? Which months had almost the same number of hours? Think about how this data could be shown in a bar graph. How would you change the graph?

# Circle Graphs

Circle graphs use a circle shape. Parts of the circle make up a whole. These graphs are also called pie graphs. Each part looks like a pie slice. You can do a tally and use that data to create a circle graph.

## Plants in Our School Garden

**1** Record your data. This data tells about plants in a school garden.

| Plants without flowers |卌 l | 6 |
|---|---|---|
| Plants with blue flowers | lll | 3 |
| Plants with yellow flowers | ll | 2 |
| Plants with white flowers | l | 1 |

**2** How many pieces of data do you have? Begin with a full circle and create the parts.

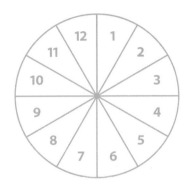

**3** Each colored part of the circle shows a value. Which type of plant has the greatest number? How can you tell? Think about the data in this graph. How could you create a pictograph with this data?

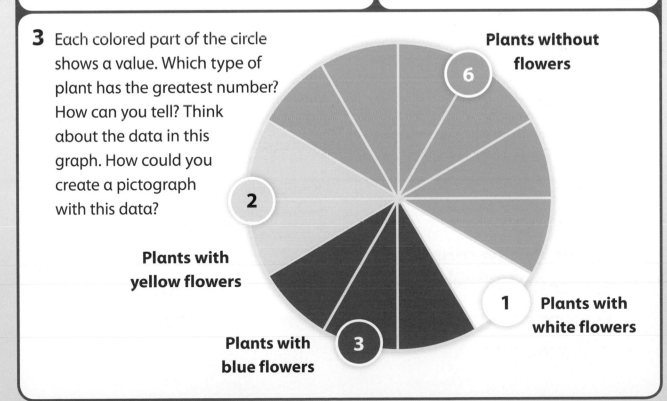

Plants without flowers

Plants with yellow flowers

Plants with blue flowers

Plants with white flowers

# Glossary

## C

**clear** (KLEAR)
A clear object does not block any light. You can see through it. (p. 162)

**communicate** (kuh-MYŪ-ni-kāt)
When you communicate, you pass information from one person to another. (p. 172)

**compare** (kum-PAIR)
When you compare, you tell how objects are alike and different. (p. 34)

**column** (KO-lum)
A column is a vertical section of a table. (p. 186)

## D

**data** (DĀT-a)
Data are observations and information that are collected and recorded. (p. 12)

**design** (di-ZĪN)
You design something when you make a plan, which may include a sketch or model. (p. 68)

## E

**evidence** (EV-i-dens)
A piece of evidence is an observation that supports an idea or conclusion. (p. 13)

## F

**fair test** (FAIR TEST)
In a fair test, you change only one thing in an investigation and keep everything else the same. (p. 13)

**flower** (FLOW-ur)
A flower is the part of a plant that makes fruits and seeds. (p. 26)

**fruit** (FRŪT)
Fruit is the part of a plant that contains the seeds. (p. 26)

## G

**grasp** (GRASP)
When you grasp something, you pick it up and hold it tightly. (p. 48)

## I

**infer** (in-FUR)
When you infer, you use what you know and what you observe to make an explanation. (p. 10)

**investigate** (in-VES-ti-gāt)
You investigate when you carry out a plan to answer a question. (p. 12)

## L

**leaves** (LĒVZ)
Leaves are parts of a plant that use light and air to make food. (p. 24)

**life cycle** (LĪF SĪ-kul)
The stages a living thing goes through make up its life cycle. (p. 32)

**light** (LĪT)
Light is something that makes it possible to see. (p. 156)

## M

**model** (MO-del)
In science, models are used to explain or make predictions about an event you observe. A model can show how a process works in real life. (p. 10)

**moon** (MŪN)
The moon is the natural satellite of Earth. (p. 106)

The **flower** will form seeds.

# O

**observe** (ub-ZURV)
When you observe, you use your senses to gather information about an object or event. (p. 10)

# P

**pattern** (PA-turn)
A pattern is something that repeats over and over again. (p. 32)

**protect** (prō-TEKT)
To protect is to prevent someone or something from getting hurt. (p. 50)

# R

**reflect** (ri-FLEKT)
An object that reflects light makes the light bounce back. (p. 168)

**respond** (ri-SPAHND)
To respond is to react to something. (p. 30)

**root** (RŪT)
A root is the part of a plant that takes in water and helps hold the plant in place. (p. 25)

**row** (RŌ)
A row is a horizontal section of a table. (p. 186)

# S

**season** (SĒ-zen)
A season is a division of the year, such as winter, spring, summer, or fall. (p. 126)

**seed** (SĒD)
A seed is the part of a plant from which another plant can grow. (p. 26)

**seedling** (SĒD-ling)
A young plant that is grown from a seed is a seedling. (p. 33)

**shadow** (SHA-dō)
A shadow is a dark place under or beside an object where light is blocked. (p. 166)

**sound** (SOWND)
A sound is something that is heard. (p. 142)

**star** (STAHR)
A star is an object in the sky that gives off light and heat. (p. 98)

The camels block the sunlight and make a **shadow** on the desert sand.

**stem** (STEM)
The stem is the part of a plant that carries water and food to the leaves and food back to the roots. (p. 25)

**sun** (SUN)
The sun is the star that is nearest to Earth. It gives off light and heat. It can be seen during the day. (p. 98)

**sunrise** (SUN-rīz)
Sunrise is the time in the morning when the sun appears on the horizon. (p. 128)

**sunset** (SUN-set)
Sunset is the time in the evening when the sun disappears below the horizon. (p. 128)

**survive** (sur-VĪV)
To survive means to stay alive. (p. 22)

# V

**vibrate** (VĪ-brāt)
To vibrate means to move quickly back and forth. (p. 142)

# X

***x*-axis** (EKS AKSIS)
The *x*-axis on a graph is the horizontal base line. The *x*-axis is usually presented on the bottom of the graph. (p. 186)

# Y

***y*-axis** (WI AKSIS)
The *y*-axis on a graph is the vertical base line. The *y*-axis is usually presented on the left side of the graph. (p. 186)

The drummers strike the drums. The drums **vibrate.** You hear the sound.

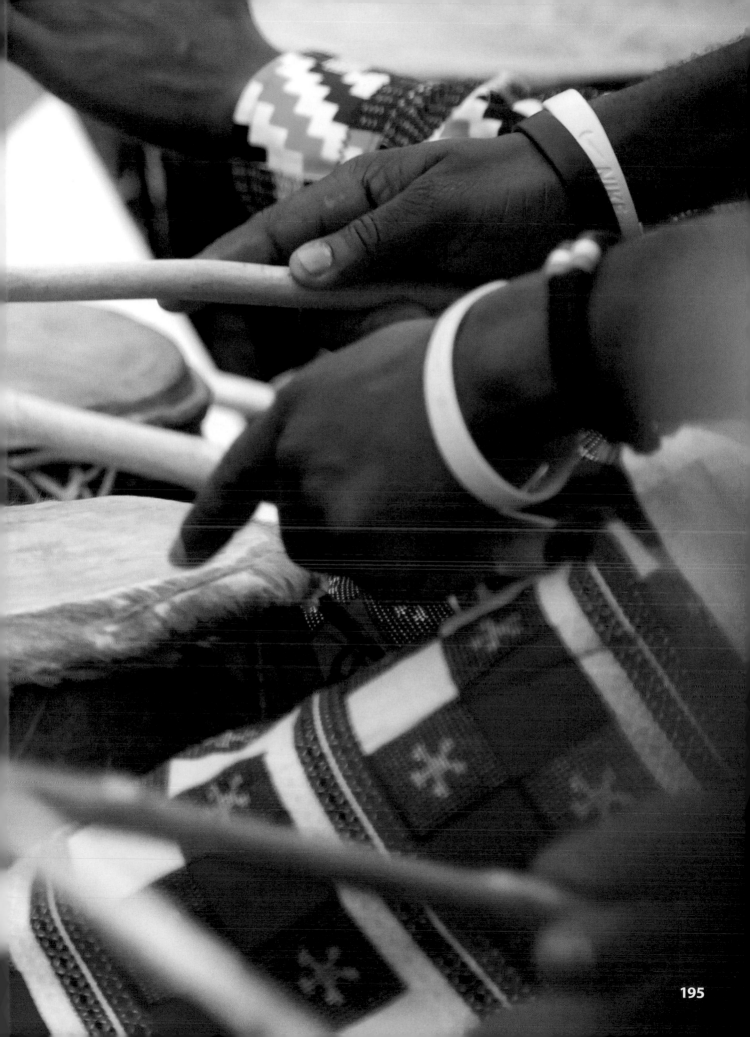

# Index

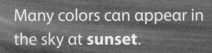

Many colors can appear in the sky at **sunset.**

The **needles** on this evergreen tree are leaves. Seeds grow in cones.

These giraffes live in a **savanna** in Africa.

This butterfly has markings of different colors and shapes on its **wings**.

135 (tr) ©Knicole Colon. 137 ©Jon Huey courtesy Nalini Nadkami. 138 (tr) ©University of Utah Communications courtesy of Nalini Nadkarni. 138-139 (c) ©Rico Ködder/EyeEm/Getty Images.

**Physical Science: Waves: Light and Sound**

140-141 (c) ©Xavier Arnau/Vetta/Getty Images. 142-143 (c) ©Jinxy Productions/Getty Images. 143 (tr) ©Michael Goss Photography/National Geographic Learning. 144 (c) ©National Geographic School Publishing. (bl) ©Michael Goss Photography/National Geographic Learning. (cl) ©National Geographic Learning. (bc) ©National Geographic School Publishing. 144-145 (c) ©moodboard/Cultura/Getty Images. 145 (tl) ©National Geographic Learning. (tr) ©Michael Goss Photography/National Geographic Learning. 146-147 (c) ©1001nights/E+/Getty Images. 148-149 (c) ©Aurora Photos, USA. 150-151 (c) ©Sue Bishop/Photolibrary/Getty Images. 152 (cl) ©Michael Goss Photography/National Geographic Learning. 152-153 (c) ©Waterfall William/Perspectives/Getty Images. 153 (tl)(tr) ©Michael Goss Photography/National Geographic Learning. 154-155 (c) ©Don Bayley/E+/Getty Images. 156-157 (c) ©myu-myu/Flickr/Getty Images. 158-159 (c) ©Karen Doody/Stocktrek Images/Getty Images. 160 (cl) ©Michael Goss Photography/National Geographic Learning. 160 (c) (cr) ©National Geographic Learning. 160-161 (c) ©MICHAEL HANSON/National Geographic Creative. 161 (tr)(tl) ©Michael Goss Photography/National Geographic Learning. 162-163 (c) © Sean White/Design Pics/Getty Images. 163 (tr) ©Marco Grob/National Geographic Creative. 164-165 (c) ©Darrell Gulin/DanitaDelimont.com. 166-167 (c) ©JASON EDWARDS/National Geographic Creative. 167 (tr) ©Dorling Kindersley/

Dorling Kindersley/Getty Images. 168-169 (c) © Frans Lanting/National Geographic Creative. 169 (tr) ©Michael Goss Photography/National Geographic Learning. 170-171 (c) ©Brian Skerry/National Geographic Creative. 172-173 (c) ©Andrea Pistolesi/The Image Bank/Getty Images. 173 (tr)(tl) ©Michael Goss Photography/National Geographic Learning. 174 (t) ©Clara Richmond/UCLA. (b) ©Felicia Ramirez/Daily Bruin/UCLA. 176 (bl)(bc) ©Michael Goss Photography/National Geographic Learning. 176-177 (c) ©Ken Welsh/Photodisc/Getty Images. 177 (tl)(bl) ©Michael Goss Photography/National Geographic Learning. 178-179 (c) © Tiffany Rose/WireImage/Getty Images. 180 (bl)(br)) ©Gabby Salazar. 180-181 (c) ©Gabby Salazar. 181 (tr) © Bill Campbell. 183 (bc) ©Jon Huey courtesy Nalini Nadkami. 184-185 (bg) ©asiseeit/Getty Images.

**End Matter**

190-191 (c) ©Barbara Friedman/Moment/Getty Images. 192-193 (c) ©Martin Harvey/Photolibrary/Getty Images. 194-195 (bg) ©klazing/iStock/Getty Images Plus. 196-197 HawaiiBlue/Flickr/Getty Images. 198-199 bgfoto/E+/Getty Images. 200-201 (bg) Angelo Cavalli/Corbis/Getty Images 202 (bg) Bernard Castelein/NPL/Minden Pictures.

**Illustration Credits**

Unless otherwise indicated, all illustrations were created by Lachina, and all maps were created by Mapping Specialists.

# Program Consultants

**Randy L. Bell, Ph.D.**
Associate Dean and Professor of Science Education, College of Education, Oregon State University

**Malcolm B. Butler, Ph.D.**
Professor of Science Education and Associate Director; School of Teaching, Learning and Leadership; University of Central Florida

**Kathy Cabe Trundle, Ph.D.**
Department Head and Professor, STEM Education, North Carolina State University

**Judith S. Lederman, Ph.D.**
Associate Professor and Director of Teacher Education, Illinois Institute of Technology

**Center for the Advancement of Science in Space, Inc.**
Melbourne, Florida

---

### Acknowledgments

Grateful acknowledgment is given to the authors, artists, photographers, museums, publishers, and agents for permission to reprint copyrighted material. Every effort has been made to secure the appropriate permission. If any omissions have been made or if corrections are required, please contact the Publisher.

 is a registered trademark of Achieve. Neither Achieve nor the lead states and partners that developed the Next Generation Science Standards was involved in the production of, and does not endorse, this product.

### Photographic and Illustrator Credits

**Front cover wrap** ©Suzi Eszterhas/Minden Pictures
**Back cover** ©University of Utah Communications courtesy of Nalini Nadkarni

Acknowledgments and credits continue on page 203.

For product information and technology assistance, contact us at Customer & Sales Support, 888-915-3276
For permission to use material from this text or product, submit all requests online at **www.cengage.com/permissions**
Further permissions questions can be emailed to **permissionrequest@cengage.com**

National Geographic Learning | Cengage
1 N. State Street, Suite 900
Chicago, IL 60602

National Geographic Learning, a Cengage company, is a provider of quality core and supplemental educational materials for the PreK-12, adult education, and ELT markets. Cengage is a leading provider of customized learning solutions with employees residing in nearly 40 different countries and sales in more than 125 countries around the world. Find your local representative at **NGL.Cengage.com/RepFinder**.

Visit National Geographic Learning online at **NGL.Cengage.com/school**

ISBN: 978-13379-11641

Printed in the United States of America
Print Number: 02
Print Year: 2019